T0092867

THE WEIGHT OF NATURE

THE WEIGHT

HOW A CHANGING CLIMATE

CLAYTON PAGE ALDERN

OF NATURE

CHANGES OUR BRAINS

DUTTON

DUTTON

An imprint of Penguin Random House LLC
penguinrandomhouse.com

Excerpt on page ix from "Landscape with Fruit Rot and Millipede" in *War of the Foxes*.
Copyright © 2015 by Richard Siken. Reprinted with permission of The Permissions
Company, LLC, on behalf of Copper Canyon Press, coppercanyonpress.org.

Excerpt on pages 86–87 from Douglas T. Kenrick and Steven W. MacFarlane, "Ambient
Temperature and Horn Honking: A Field Study of the Heat/Aggression Relationship"
in *Environment and Behavior* 18, no. 2, 179–191. Copyright © 1986 by SAGE
Publications. Reprinted by permission of Sage Publications.

Sections of Chapter 8 were adapted from Clayton P. Aldern, "Mountaintop Removal
Country's Mental Health Crisis." Copyright © 2016 by Grist. Reprinted by permission of
Grist, grist.org.

Additional material in Chapter 8 previously appeared in *It's Freezing in LA!* and *Crosscut*
and was adapted here for publication.

LIBRARY OF CONGRESS CATALOGING-IN-PUBLICATION DATA

Names: Aldern, Clayton Page, 1990– author.
Title: The weight of nature: how a changing climate changes our brains /
    Clayton Page Aldern.
Description: New York : Dutton, 2024. | Includes bibliographical references and index.
Identifiers: LCCN 2023048201 | ISBN 9780593472743 (hardcover) |
    ISBN 9780593472767 (ebook)
Subjects: LCSH: Human beings—Effect of climate on. |
    Climatic changes—Health aspects. | Brain—Evolution.
Classification: LCC GF71 .A478 2024 | DDC 304.2/8—dc23/eng/20240131
LC record available at https://lccn.loc.gov/2023048201

Printed in the United States of America
1st Printing

BOOK DESIGN BY LAURA K. CORLESS

Interior art: brainlike tree branches © Adrian Picon / shutterstock.com

For Ally,
a light of nature

CONTENTS

PART III: DISPLACEMENT
Sensing, Pain, Language

. . . The mind fights the

body and the body fights the land. It wants our bodies,
the landscape does, and everyone runs the risk of
being swallowed up. Can we love nature for what it
really is: predatory? We do not walk through a passive
landscape. The paint dries eventually. The bodies

decompose eventually. We collide with place, which
is another name for God, and limp away with a
permanent injury. . . .

—Richard Siken, "Landscape with Fruit Rot and Millipede"

THE WEIGHT OF NATURE

TENSION

I will make justice the measuring line and righteousness the level. Hail will sweep away your refuge of lies, and water will flood your hiding place.
    —Isaiah 28:17

You are not surprised at the force of the storm—
you have seen it growing.
    —Rainer Maria Rilke, "Onto a Vast Plain," 1905

One of the first things Dezaraye Bagalayos told me was that she felt transparent. It's not that people were seeing through her, she explained, but that she was seeing through herself. "For the last ten years, I've been feeling like I don't fully exist," she wrote. "Like my feet are never firmly on the ground." It was a liminal existence: a floating and flickering, more suited to a ghost than to a living, breathing woman with a six-year-old daughter to raise.

Dezaraye's transparency stems from what she calls her parallel lives. The first of these, the one everyone else sees, is framed by those little Lego blocks of progress: getting a college degree, advancing her career, buying a home and fixing it up, teaching Eleanor to read. She stresses the importance of education to her daughter, and they do their homework together every night: mom toward her bachelor's, Eleanor for first grade. It is a life spent working on land and water issues in California's San Joaquin Valley, a mitten-shaped expanse of farmland so lush and fertile it was once referred to as the breadbasket of the world. If you're living in the United States and have ever eaten a grape or an almond, it likely came from the San Joaquin, and it was likely watered by the Sacramento–San Joaquin River Delta.

The delta: For seven hundred meandering, dendritic miles, the waterscape fans across California like the searching roots of a peach tree. Having crept its way inland near the end of the last glacial period, the delta now soaks the crops of the San Joaquin via a series of irrigation superhighways modulated by pumps at its south end. Yet despite this imposing network of canals and aqueducts, even trillions of gallons of water aren't enough to quench the thirst of the billions of dollars' worth of Bing cherries and White Rose potatoes and Tulare walnuts rooted in the valley. In times of drought, the gap yawns wider. To make up for meager levels of surface water, farmers drill down—as deep as two-thirds of a mile—to tap Joaquin's aquifers.

If any cycle can be said to have a starting point, a candidate beginning for the vicious spiraling of the San Joaquin Valley would be here, in the depths of the groundwater wells. As farmers haul acre-foot after acre-foot of water from the aquifers to keep pace with the appetites of their crops, the damp clay and silt beds below the surface drain into newly depressurized sand and gravel. Wrung like kitchen sponges, the silt beds compress, and the bloated land above obediently settles. With a slow and silent scream, the San Joaquin Valley is sinking.

It goes like this: Record-breaking droughts slash the supply of surface water, trickling canals breed a need for deeper wells, over-drafted aquifers pull the planet's surface closer to its core—and a subsiding, buckling Earth damages the laughably fragile human infrastructure placed upon it, including the canals carrying surface water from the delta. As the cycle repeats, it turns poisonous. We wring out old agricultural land, and the compression forces arsenic into the water table. And then there is the great slow force we always believe to be operating at the margins, which here in Central California is turning the dry seasons increasingly more punishing. Some parts of the valley are now cratering at rates as high as two feet per year; others have sunk thirty feet since the 1920s. Our augers twist, and the land follows us down.

Dezaraye's other life is one pulled to this sunken place. "My daughter has no future," she says. "Our days on this planet are numbered, I don't think how we will all die will be pleasant, and nothing and no one is moving fast enough to stop it." This is the chaos of climate change. She cites the advent of water markets as paving the way for the full-on privatization of water in California; for private militias protecting the liquid's owners; for all-out resource wars in her hometown of Stockton. What more is a ghost than something caught between life and death? "I feel like how I imagine the Earth to be feeling," she says.

The story of your life is a plumb line. Which is to say: It has direction. You feel this direction, because you point yourself in it every moment of every day. You decide to get out of bed, eventually. You

3

decide to brush your teeth—or you decide to skip. Sometimes you decide to crack open a book beside your banana with coffee; sometimes you decide to continue reading to the next paragraph.

The direction is not wholly yours.

It was hot last night, and you didn't sleep very well. You are groggy today, which is to say you are acting a certain way because you are feeling a certain way. That banana is smiling at you a little funny, and it takes two cups of coffee for you to feel a sense of control again. Sip, sip. Okay. You are becoming yourself.

What happened is the world did something to you. It pulled at you, just a little, and you tilted off your course, just the smallest bit. It was barely at all: the flap of a butterfly's wings in the august maelstrom of life. But look—what we're talking about here is the opposite of the butterfly effect. There is no marvelous domino amplification of sneezes into tidal waves. No, all we care about for now are the little changes that follow the other little changes. It was hot last night, and here you are, reaching for a little more caffeine. And yes, it is still *you* who's reaching—the world's not doing that. You still get to make decisions when the world pulls you off course. What I mean is: Your storyline still has direction. But there is something else in the room. There is a weight tied to the line of your life, and it points toward Earth.

This is a book about your second cup of coffee. It is about the ways in which the natural world tugs and prods at the decisions you make; how it twists and folds your memories and mental states; how this nebulous everywhere we call the environment tips your interior scales. Sometimes these nudges are benign: You snap at your mother on the phone because surely you have had this conversation before, and right now it is 95 degrees Fahrenheit (35°C) and your patience has run dry. Or maybe it's the air quality that's off-kilter today—there's that wildfire across the state, after all—and you have an absolute whopper of a headache, and you simply cannot focus enough to remember if your partner's birthday is January 5 or January 6. You should absolutely know the answer to this question. Is it really the same day as the Capitol Hill riot? That seems too on the nose.

And it is not so bad to forget a birthday or snap at your mom. You can remedy those kinds of things, even when it's too hot to sleep. But that is not always the case. Sometimes the world offers more than a nudge: Sometimes it pushes further, and sometimes you are changed forever. The ragged trauma of a hurricane, wildfires that melt people—the interior demons that check in to the motel of your mind and never leave, and somehow convince you to pick up the tab. There is the abyssal depression that descends with landscape loss; the cortex-rotting diseases provoked by warming waters. There are, with us now, brain-eating amoebae and plummeting test scores and the shrunken brains of chronic stress. All this and more. What I am trying to say is: The horrors are not out there. The horrors are inside.

But let's start with coffee. We're still waking up.

For millennia, people have understood a certain limitlessness of humankind. You can't get past the first chapter of Genesis without being clubbed with the notion that man was given dominion over all the Earth. Omnipotence and agency and perfectibility are baked into our understanding of what it means to be us. Aspens, gazelles, alfalfa, carpet beetles, river pebbles, waterfalls—they all do more or less the same thing: react. But we do something very different. In history books, our greatest advances as a species are predicated on the pushing around of nature. We call the domestication of wheat an agricultural revolution, the burning of coal an industrial revolution. We are in control.

Even as I write this sentence, I can't shake the sense that it's *me* writing it, that I am the one in charge. I think, therefore I am; that old rationalist mantra. René Descartes, the French philosopher who gave us the idea, thought free will was self-evident—that "it must be counted among the first and most common notions that are innate in us." In fact, for Descartes, free will not only separated us from nature, but also created this dualism between our minds and our bodies. In an otherwise predetermined world, our thoughts defied physics. They could move limbs. They could have a second cup of coffee. Cartesian dualism protected the mind from the influence of the environment. It didn't matter how the Earth felt: We were the only ones with feelings.

As it would turn out, Descartes was no neuroscientist. Nobody was in 1644, when he was penning his treatises on free will. Over the next 375 years, though, careful thinkers—scientists, philosophers, clergy, artists—learned much about the brain. We learned it was made of tissue, of cells, just like the rest of our bodies. It was indeed the seat of thought, but it didn't defy physics at all. We learned that these brain cells communicate with one another in a language of chemicals and electricity; that this language undergirds our own—spoken, written—as well as our movements and emotions and sense of self. Neuroscientists are still working to crack the code of this electro-chemical babble. It is a subtle, complicated art, decoding exactly *how* the brain does what it does. But over four centuries, we have assembled a better idea of *what* it does and doesn't do.

When I worked as a neuroscientist, that's what I thought I was supposed to be doing, anyway. I poked at computers, mostly, modeling the brain and its circuits. By constructing computer models of neural activity, my colleagues and I could make predictions about what was happening in the messy, wet, real deal as the organ went about its business. In theory, our efforts made the job of the experimentalists— the people who worked with real brains—easier. We built models to generate hypotheses about what the brain was doing, the people who stuck electrodes into real brains collected neural data and compared it to our models' predictions, we adjusted our models as a function of any mismatches; rinse, repeat. Eventually, so went the thinking, we'd be left with a reliable model of the brain that captures how the real thing behaves. And then you can use this model to make predictions about, say, the potential efficacy of certain pharmaceutical therapies. You can perturb your model to understand what's happening in the case of brain diseases.

We theorists—the modelers—we were a little bit *like* the brain, actually. It turns out that what the brain does, broadly, is help instantiate a model of the world. That's because in order for you to navigate your environment, you have to possess an innate sense of how it all fits together. If you're to survive out there in the concrete jungle, gravity can't surprise you. You need to understand that when airplanes disappear behind clouds, they're not going away forever. You need to remember what happens when you move your feet forward. Your brain helps build and store these kinds of predictions about the ways of the world. These predictions are within you. In other words, you *are* a model. You are a picture of what's out there.

But the picture isn't static. From moment to moment, to sustain your existence, your brain compares the predictions of its world model to the sensory information it receives, tweaking its inner workings in order to minimize surprise. You look around, feel about, move hither and thither—all the while expecting things to go a certain way. When they don't, you update your model accordingly. As with the interplay between theoretical and experimental neuroscientists, the goal here is to *minimize the mismatch between what you expect to experience at any given moment and what you actually experience*, given the sensory impressions the world grants your eyes and ears and other sensory systems. It's the only way you can exist continually in time. If your brain didn't seek to minimize surprise, you'd be pathologically dumbstruck, every moment of every day. You would forget that people generally have two arms; you would be terrified to learn your hands are attached to your body and that the sky is such a remarkable shade of blue. But instead of surfacing a constant nightmare, your model learns to expect these kinds of things so it can focus on the interesting stuff. Every time you look around a room, you are updating your model of the world—your awareness of it and your place within it. Modeling the world allows us to understand that we are still alive and that reality looks roughly the way we expect it to look. Our conscious access to this model—our feelings and our knowledge—allows us to use our brains and bodies as tools to sustain themselves.

In other words, one way to think about *feelings*—bodily sensations, the emotions we assign to them—is to understand that they offer us experiences of our own life. As the neuroscientist Antonio Damasio has written, feelings "provide the owner organism with a scaled assessment of its relative success at *living*." Much ink has been spilled in service of the notion of regulating our emotions. It's all roughly backward. The thing we were designed to do is let our emotions regulate us. Success at living means listening to your body.

An understanding of yourself as a model builder necessarily invokes the brain *and* the rest of the body. The brain is useless without its home. Without being able to see or touch or smell the world, without being able to navigate it with your legs or, say, with your arms in a chair, without knowledge from your organs as to whether you're hungry or fearful or burning yourself, your brain wouldn't have anything to model. Which is to say: One thing the brain *doesn't* do is operate in isolation, and this is why Descartes was so wrong. Counter to the instincts of the dualists, contemporary neuroscience has taught us of the inextricability of the mind and the body. Cognition is literally *embodied*. The conscious, curious mind may indeed arise from some unassuming wrinkly organ tucked away beneath a few millimeters of skin and skull and spidery protective tissue—but just as fundamentally, this sorcery depends on the interactions between this organ and the rest of the animal in which it sits. The stuff of thought is physical stuff. It is exposed to the world, and it makes itself in its image. In your second cup of coffee is a strong shot of reciprocity.

Embodied cognition implies that the mind is subject to the whims of a wild planet. You mirror your environment, and not in an esoteric Age of Aquarius sense. Much of what I suggest in this book can be boiled down to the following: As the environment changes, you should expect to change too. It is the job of your brain to model the world as it is. And the world is mutating.

For the past several years, I've been talking to people who are feeling like the Earth: uncertain, oscillating, strained to the limit. A weary world is wearing on us, and as climate change forces the seas and ice and heat index to their extremes, the extremities reach back. We have words for these conditions now: climate grief, ecoanxiety, environmental melancholia, pre-traumatic stress disorder—a whole new lexicon to describe the psychological and emotional chaos of coming to know our degrading planet more intimately. High schoolers are preparing for a broken climate future with the same urgency, fear, and uncanny resignation they usually reserve for active-shooter drills. There are some people, like Dezaraye, who worry that their children have no future; there are others who may not have children at all for fear of introducing them to a burning world. How many of us share this dread?

I'd wondered as much. It didn't take long to find others: more of us, bending under the weight of the new normal. Parents and grandparents who have watched timeless shorelines erode and billows of insects vanish and the four seasons meld into one; their children, a younger generation who have only ever grown up under climate disruption, now coming to understand the unfairness of the cards they've been dealt. In 2015, psychologists introduced me to the burgeoning field of climate psychology. In 2017, economists pointed me to the ever-stickier fingerprints of the environment on human behavior. And then, just as my research began to ring with a familiar, confirmatory echo—you are undoubtedly by now familiar with the notion of "climate anxiety"—something happened: I began to speak with my former colleagues, the neuroscientists, and I realized they were scaring me. The crux of the relationship between environment and mind sat deeper than I'd ever imagined. Climate change wasn't only here; it was inside us.

In a jarring paradigm shift, we're watching a rapidly changing environment directly intervene in our brain health, behavior, cognition, and decision-making in real time. Temperature spikes drive surges in everything from aggravated assault to domestic violence to online hate speech. Surging carbon dioxide levels and heat waves diminish problem-solving abilities, cognitive performance, and our capacity to learn. You don't need to go to war to suffer from posttraumatic stress disorder: The violence of a wildfire or hurricane, now crashing down with ever-increasing frequency, will do the trick.

Climate-fueled neurotoxin exposures; climate-fueled brain diseases. And then there is our leviathan of climate grief. From anxiety to productivity to fear, memory, language, the formation of identity, and even the structure of the brain, the forces of the natural world are there, exerting an invisible but unmistakable nudge—nature's thumb on the scale of our inner workings. It's all part and parcel of a simple, striking revelation: We haven't been the only ones pushing. Nature pushes back, and it does not hesitate to shove.

This book is about the neuroscience and psychology of how a changing world changes us from the inside out. It is not a book about climate anxiety—about *worrying* about climate change—though we'll touch briefly on the seriousness of the subject later on. It is also not a book about climate communications or climate politics and the manners in which the psychological dimensions of the climate problem *lend themselves* to complicating those arenas. That's important stuff; you can read about it elsewhere. I also won't wade too deeply into the trenches of consciousness, though a useful reminder-mantra might be: The mind is grounded in the brain and the body. Mental energy is physical energy. There; that's about all you're going to hear from me on neurophilosophy. In this book, instead, you'll only find evidence of direct interventions of environmental change on the brain and mind. That is the scope of this thing.

Here is who I am and why I care about any of this mess. These days, I often introduce myself as a recovering neuroscientist. Back in 2015, when I first conceived this project, I had just finished a short graduate program in neuroscience at the University of Oxford, and I was spending much of my time fiddling with a programming language called MATLAB, trying to replicate in a computer the results of an experiment someone had conducted on zebra-finch brains. I was known as a laboratory technician. I worked in a place called the Centre for Neural Circuits and Behaviour. I'd thought I'd landed my dream job. I was concurrently enrolled in a graduate degree in public policy. I drank cask-temperature ale and cold Pimm's cups and played rugby and puttered around in boats called punts and occasionally wore a formal cape. Charmed life. I'm still grateful today.

To be blunt, though, it all felt a little disconnected from the world that existed outside the Centre.

And then a friend showed me a report from the Pentagon. Earlier that year, the Department of Defense had quietly submitted a slim report to Congress on the national security implications of a changing climate. The fourteen-page document, written in the well-oiled, acronym-heavy prose of the military apparatus, was remarkable in what it laid bare. Not only did Defense—not exactly Greenpeace— see climate change as a serious threat to national security; it noted that these effects were already occurring. The mid-2000s Syrian drought, in its choking of agriculture and flame-fanning of mass displacement, had helped spark the Syrian civil war. In 2012, on home turf, when Sandy crashed those torrents of stormwater through the hallways of New York City, the Defense Department was forced to mobilize twenty-four thousand personnel. To the military and intelligence brass, climate change wasn't hypothetical. We were already fighting it.

Most astounding in the July 2015 report, though, was a framing of *how* a changing climate acted on our security. For the Department of Defense, climate change wasn't about rising temperatures and sea-level rise. Those concerns were secondary to something more human: the global climate's ability to "aggravate existing problems" like poverty, social tensions, ineffective leadership, and weak political institutions. It wasn't just that a warmer world would hurt us outright; it was that a warmer world would make us hurt one another. I remember a jigsaw piece snapping into place in my head.

A few months before the submission of the Pentagon report, a Stanford economist named Marshall Burke and two of his colleagues had published an article in the *Annual Review of Economics* about the relationships between climate and conflict. Surveying dozens of independent studies, the team illustrated how deflections in temperature or rainfall patterns were associated with increases in both large-scale conflicts and individual violent crimes. The extraordinary finding here, surfacing again and again in the studies Burke's team reviewed, was that these crime spikes couldn't be pinned to things like income or food insecurity. The statisticians could correct for all those effects. There was something else going on.

It's the kind of finding that sets off alarm bells in the ears of neuroscientists. With a Stanford professor and now the Department of Defense claiming temperature changes were somehow intimately related to violence, I began to wonder which tangle of neurological pathways might be at work—which strands, if any, could be separated from the geopolitics and the sociology and the anthropology—and what this rat's nest might mean for a warming world. Climate change was always out there, in that other place and future time. If droughts and rising temperatures were doing something to people now, in *here*, that most personal of spaces, it seemed worth asking how and why. I channeled my policy degree into climate policy; I left the lab for environmental journalism. This book is the product of the seven years of research that followed.

As suggested earlier, the link between temperature and aggression is the tip of a melting iceberg. As I dug into the neuroscience of environmental change, evidence of invisible forces emerged by the truckload. A changing Earth was infiltrating our internal worlds, guiding our hands, and putting words in our mouths. The environment was everywhere. And of course it was. I was a fish coming to understand water for the first time.

Together, the chapters presented here represent a fundamental piece to the environmental puzzle that in our climate modeling, policymaking, and personal lives we are broadly ignoring—to our own detriment. The effects of climate change on our brains constitute a public health crisis that has gone largely unreported. And it is past time to act. Immigration judges are more likely to reject asylum applications on hotter days. Some drugs that act on the brain aren't as effective at higher temperatures. Wildfires evict us; chronic stress becomes clinical. In the world of disease vectors, climate-fueled ecosystem changes are expanding the reach of everything from the mosquitoes of cerebral malaria fame to brain-eating amoebae most of us have no reason to have ever even heard of. Major depression symptomatology skyrockets with landscape loss. Your high schooler is going to lose a couple of points on the SAT if they take it on a particularly hot day. We are already victims of the climate crisis, whether we know it or not.

Which is frightening. It should be, anyway. It frightened me when I began to understand it, and it frightens me now.

I don't seek to leave you in the fetal position, though. And I don't think that needs to be our response. I feel this way for a couple of reasons. The first is that some of our impending public health nightmare is not without solutions. Some are technical in nature: infectious disease protocols, pharmacological fixes, stopgaps like air-conditioning and air-filtration systems. But much of the psychological, mental, and emotional resilience here will be rooted in techniques adapted from psychotherapy and the behavioral sciences—adaptation responses we can all apply to creatively channel grief, avoid burnout, reclaim agency, and find purpose in the face of an arbitrary, unjust future. *Feeling* a changing climate may be the means by which we ultimately respond to it. Collective action can foster connection and grit. We have storytelling too: that ancient means by which we acknowledge and name our experiences, enter the minds of one another, and hone the imperfect art of empathy. There are some unifying, connective principles here. They can ground us.

But first we have to pay attention.

In a 2016 interview with *The New York Times*, Barack Obama called climate change "a slow-moving issue that, on a day-to-day basis, people don't experience and don't see." Indeed, until recently, most of us tended to read climate change as an abstract idea: this slow, plodding thing inching across borders and generations. We only paused to really think about it when that plodding was punctuated by those aberrant hurricanes and hundred-year wildfires. Just a few short years after Obama's interview, it's no longer true: We see evidence of climate devastation in headlines, in photographs, and through our windows every day. Environmental change is no longer just in our periphery. Philadelphia's summer is like Atlanta's used to be just a few decades ago. Atlanta, for its part, now has summers like Tampa's. Recent wildfire seasons have illustrated in drastic orange and black the roaring consequences of an untamed climate. Hurricanes are swallowing coastal neighborhoods. July 2021 saw simultaneous extreme weather on four continents, from lethal flooding in Western Europe to raging forest fires in Siberia to China's heaviest rain in a thousand years.

There's more to see here than extreme weather, though—if we know where to look. The neurological, emotional, and behavioral effects of environmental degradation can pull our understanding of climate change out of the future and into the present. The climate has changed, it continues to do so, and millions of people are already facing the consequences. By centering human experience as it relates to a changing climate, we can give tangible contours to the formless. The twisted, humbling paradox of it all is that we perhaps only ever needed to look inside ourselves.

Doing so today can help us better conceptualize this crisis—the greatest existential crisis humanity has ever faced—acknowledge and foster resilience in our responses to it, and lift up viable solutions for navigating our new normal. In short, we can grasp the importance of global environmental change by more deeply considering ourselves. Social scientists tell us that climate change demands emotional engagement if policymakers are to listen and act. I think that's almost right. People can and ought to connect with a changing climate on emotional terms—but if the climate is already changing us, we already do.

There is a certain kind of introduction to a certain kind of book written over the course of the coronavirus pandemic. It goes something like: *And then the world changed, and we recognized the degree to which everything was interconnected—how your health influenced mine; how structural power dynamics shaped access and opportunity—and we realized this project wasn't just a collection of pirate-themed cocktail recipes; it was also a history of empire and white supremacy.*

This book, the one in your hands, does not profess much moral clarity, nor has its thrust changed much over the course of the pandemic. It is a project predicated on interconnectedness, but to state as much feels like reminding someone that their feet are on the ground because of gravity. It's axiomatic enough to be uninteresting.

It's true, though, that much of the writing here took place during a time of great consequence and mass death. It feels worth naming as much. Under pandemic waves—the curves we needed to flatten—the world became smaller. The waves pressed. They threatened to flatten *us*.

There is another wave coming. This book is sympathetic to the idea that the climate crisis is going to get worse before it gets better—and that we are drastically underestimating (and failing to articulate) many of the risks therein. But I don't think this story has to be exclusively one of fear. You can think of the following chapters as insurance brokers: They say, *Look, here are the risks we ought to be enumerating and hedging against.* We know that climate change bears on memory systems and cognition and impulsive aggression. We know it infuses the water and air with neurotoxins; that it increases the range of brain-disease vectors; that extreme weather can spur post-traumatic stress disorder. We know that a changing climate can act intimately on our sensory systems. It can seed anxiety and depression. It can corrupt language and, by extension, our perception of reality. The waves of the climate crisis are already bearing down, whether we accept it as fact or not. The question is: How will we stay above water?

I submit, perhaps a bit too earnestly for a book I'm encouraging you to take seriously, that we have to keep each other afloat. There is room on the lifeboat. Much of this text assumes an understanding of the world reaching inward. But we can reach out too, against the weight of the wave, and not just in a blind grope. People caused the climate crisis and its resulting neurological effects, and it will be people, in solidarity, who reverse it. It will be people who foster resilience in one another. You will find examples of this resilience in these pages.

The text is split into three parts. The first focuses on the influential action of climate change on behavior: its sapping of our perceived free will through the lens of memory systems, cognition, and aggression. The second considers the manners in which a changing environment enters explicitly into friction with neurological health. I cover environmental neurotoxin exposures bolstered by climate change, climate-fueled infectious brain diseases, and extreme mental trauma. The third and final section concerns subtler actions. It is about how climate change and environmental degradation infiltrate sensory systems like olfaction and audition; how climate change influences pain communication and major depression; and how it insidiously shapes language and perception. It is an unnerving, heavy pile of evidence.

But this book is not the wave. It is a hand reaching out. Hold on.

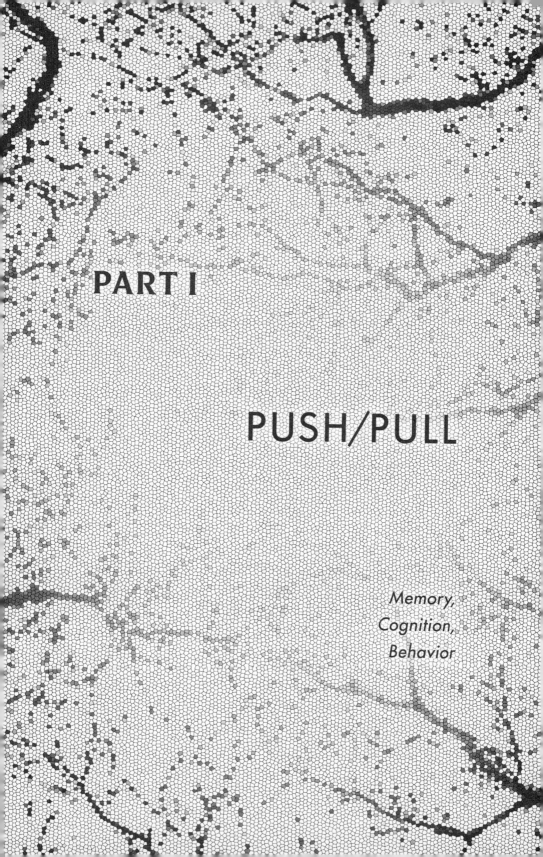

PART I

PUSH/PULL

Memory,
Cognition,
Behavior

1

A HISTORY OF FORGETTING

If we could not forget, we could not remember; just as only the trembling balance can weigh.
—Erwin Chargaff, *Heraclitean Fire*, 1978

I will wonder if the memories that remain with age are heavier than the ones we forget because they mean more to us, or if our bodies, like our nation, eventually purge memories we never wanted to be true.
—Kiese Laymon, *Heavy*, 2018

To the extent that any year can be dehydrated down to a phrase, 1896 was the year of detection. On New Year's Day, the German physicist Wilhelm Röntgen wrote in a letter to colleagues that he had been able to produce a kind of "shadowgram" of his wife's hand—he showed his friends how you could see Anna's bones in the print. There was some flavor of radiation involved; Röntgen couldn't quite put his finger on it; he called the rays "X." That same year,

Tagish and settler prospectors struck gold in a Klondike creek that would go on to be renamed Bonanza. There was the French physicist who reported that uranium emitted its own variety of invisible rays, all by itself, in the dark. And in April, the Swedish scientist Svante Arrhenius published a quiet little study illustrating how the burning of fossil fuels could release carbon dioxide into Earth's atmosphere, and how this gas could remain there and trap heat, and how this effect could warm the planet's surface "like the glass of a hothouse."

A thousand articles on X-rays would be published in 1896 alone; some hundred thousand prospectors would set off for the Yukon after Keish and the Carmacks disinterred its gold. But Arrhenius's musings on coal-fueled planetary change would collect dust on a shelf while industrialists burned dead things for another century.

Not that we didn't notice the warming. A few decades later, on September 29, 1933, a US Weather Bureau chief named Joseph Burton Kincer sent an essay along to the *Monthly Weather Review*, posing an honest question in its title: "Is Our Climate Changing?"

Kincer was puzzled by his instruments. Frankly, he was puzzled by the previous thirty years. The cool periods were coming less frequently, and they didn't stick around as long; the warms were more pronounced and persistent. He couldn't spin it any other way. "Historic climate has always been considered by meteorologists and climatologists to be a rather stable thing," he mused—the climate was not the same as the weather. And yet there the numbers were. Nervous energy: a shuffle of data sheets, a flick at the thermometers. A new climatological regime? The stuff of heretics. But the contradiction remained, and so Kincer put pen to paper. He wrote of the "apparent longer-time change" the previous decades appeared to herald. He wrote of the changing of the seasons. His tone was strained because he didn't want to give up on predictability. Predictability was what he had.

He wrote of his grandfather:

It appears, however, from the data presented with this study that the orthodox conception of the stability of climate needs revision, and that our granddad was not so far wrong, as we have been wont to believe, in his statements about the exit of the old-fashioned winter of his boyhood days. We are familiar with statements by elderly people, such as "The winters were colder and the snows deeper when I was a youngster," and the like.

Kincer wasn't about to deny the evidence before him, but he didn't mind an argument with Granddad. And surely the winters would come back. In his essay, the scientist made clear he believed the world had found itself in some kind of protracted cycle as opposed to slingshotting along any runaway trajectory. "Is our climate changing?" he asked. It would seem that way, but just for a time. Kincer was chief of the Division of Climate and Crop Weather; the stability of the climate underwrote his salary.

He would still be clinging to the static picture of the seasons in 1939, when he argued in another essay that the span of a few decades simply wasn't enough to establish an understanding of normality—you needed to go back further. A peek at the previous *century*, for example, indicated that recent warming did "not represent a permanent change of climate, but rather a warm, dry phase of our normal climate, to be followed, doubtless, by a cooler, wetter phase, when there will be more rain in summer and lower temperatures in winter." Doubtless.

By 1946, two years into his retirement, whatever had begun around the turn of the twentieth century hadn't yet reversed. And though the world had been warming for more than half his life at this point, Kincer dug in. Looking back on the thirteen years following his original analysis, he concluded that "the general upward temperature trend continued for several years but that the more recent records indicate a leveling off, and even contain currently a suggestion of an impending reversal." There was still time for Granddad's winters. The climate ebbs and flows. He would die a few years later.

In Kincer's far future, on the day I copy down this last quote, Antarctica witnesses a 70-degree-Fahrenheit (21°C) deviation from normal. It is a heat wave unlike anything ever recorded: As far as scientists can tell, the event represents the farthest a weather station's sensors have ever strayed from average. Anywhere, for all time. It's the kind of record that scoffs at the notion of normality. But what else do we have if we don't have our expectations?

To notice the extraordinary, we first need to have formed a belief about the world: something like "The winter is cold" or "April showers bring May flowers." These beliefs define what it means for something to be novel. Novelty—newness—is only that which upsets our expectations. It's why card tricks are so memorable: They disrupt our internal model of the way things are supposed to work. A deck of cards doesn't *usually* hint at magic—until you show me the guillotined queen, now reconstituted. Didn't I watch you cut her in half? The sky is *usually* blue—until the fires turn it orange. It's in this manner that climate science is innately gummed up with memory and expectation. The climate can't change if you don't know what it felt like to begin with.

And yet that last sentence is a card trick, too, because something else bordering on magic just happened: Fluidly and instinctually, you accepted the premise that you can feel the climate at all. Read it again, slowly: The climate can't change if you don't know what it felt like to begin with.

What does the climate feel like? Is it velvety? Rubbery? Where does it begin and weather end?

The climate is a description. It is a cultural expectation. Often this expectation is expressed in the language of science, but it doesn't have to be. The expectation is also buried deep within us: old patterns in your brain, old knowledge in your bones. We use this description—this knowledge—to make sense of the rest of the physical world, the stuff that happens outside our bodies. But the point is: Climate is not the thing you form expectations about; it *is* expectation.

Even in climate science, the word only implies a statistical description. It is "average weather," according to the World Meteorological Organization. But an average is an idea—an abstract human concept—not something that exists in the absence of imagination, as, say, a broom does. The WMO goes on to define "average weather" as "the measurement of the mean and variability of relevant quantities of certain variables (such as temperature, precipitation or wind) over a period of time, ranging from months to thousands or millions of years." Watch the magician's hands: Climate is not temperature and precipitation; it is their measurement.

Mike Hulme is probably the person on Earth who has paid the most attention to this sleight of hand. A geographer at Cambridge University, Hulme has staked a career on climate as culture. In the wide evolutionary workshop of human history, he argues, climate was a necessary invention: "The idea of climate introduces a sense of stability or normality into what otherwise would be too chaotic and disturbing an experience of unruly and unpredictable weather." Instances of *weather* are not ideas—rain exists—but they are ephemeral. The atmosphere is in constant flux. If people are to accomplish anything at all in the physical, fluctuating world around us, he argues, we need a foothold; a base camp from which to plan for the future. "Climate is something else, hinting at a physical reality that is both more stable and more durable than the weather," writes Hulme. "Unlike the weather, climate is therefore an idea of the human mind."

Stare at it enough, and Hulme's argument doesn't feel so radical. To consider climate as an idea doesn't make it any less real—it just reminds us *where* this phenomenon plays out. Without question, hurricanes batter coastlines and flatten neighborhoods and drown whole congregations. When they kill, they do not do so abstractly. But our expectations around hurricane frequency and strength—how much death is *normal*—these expectations are products of observation and reasoning and memory. They exist within us, in cultures and brains. They are grounding. They're stubborn. And they're supposed to be. As Hulme argues, we invented climate for the sake of cultural and psychological stability. The problem is in the implication: When the climate changes, so do we.

Part of the reason the climate is knitted so completely into the fabric of our being is that weather memories, in particular, are often deeply personal—and deeply emotional. Think back to your wedding; to the first time you picked raspberries; that ice-fishing trip with your stepdad; your grandmother's funeral. You remember the weather. It's also likely the associations go both ways. Dry, biting cold brings you back to that Minnesota lake. In one of its most basic expressions, neurologically, a memory *is* associative. Its instantiation in the brain is a collection of perceptual experiences that have been wired together. Remembering the ice-fishing trip means recalling the scratch of your little mittens and the toughness of the jerky you brought along and the slushing sound of the auger and the snap of the wind—all of it together. Encountering one of these facets later in life can bring the rest of them back.

Hulme's own weather memories are fearful. At age five, as summer rains flooded his family garden, Mike would feel "a primordial fear of drowning." At six, he was "battered and bruised by a violent April hailstorm." Weather wasn't statistics. It was a core aspect of his experience, and one that today makes up an account of his past. In some sense, climate is the skeleton of memory. Seasonality provides connective tissue: ligaments that join present expectations to the past experiences that shaped them. It *means something* for winter to come. As the psychologist Trevor Harley has written, "Weather can provide a framework for accessing and structuring memory. That is, the weather plays a metacognitive role in organizing cognition."

But then the climate changes and this foundation is shaken. The idea of climate change "begins to unsettle our belief that we know how the weather of the future *should* be," writes Hulme. The italics are his, and the "should" is important. Expectation is a normative guide—it has an opinion. And expectation is primitive.

We haven't always known this. A few years after the discoveries of X-rays in Würzburg and gold in the Yukon, in 1903, a graduate student at Yale named Clara Maria Hitchcock published her doctoral thesis in *Psychological Review*. "The Psychology of Expectation" begins with a card trick in its own right. Hitchcock introduces the idea of expectation as her friends and colleagues would have tended to have thought about it: as "a very simple mental process which has the same relation to the future that memory has to the past." And then she unsettles her audience:

Suppose man could not look forward, and was obliged to walk into the future backward, as it were. What would be the meaning of his past to him under such circumstances? Could it have any other than an aesthetic significance, affording him pleasure and pain, without any possible purposive application? It is absolutely essential to assume at least a contemporaneous reference to the future in order that memory and the higher thought activities may serve any practical purpose in conscious life.

Hitchcock hated memory. Or, at least, she despised its dominance in her nascent field of psychology. To Hitchcock, memory was static and stale: a crumbling monument to a bygone past. Expectation was everything else—and it was dazzling. It was "the assurance to each soul that its life is continuous, that its existence will not end with the present moment." Calling out this stabilizing function of expectation was Clara's daring contribution to her field—and to human understanding of ourselves. In her dissertation, she elevated expectation to the same psychological pedestal onto which the miracle of memory had been placed, and then she doubled down and raised it higher. While memory was "comparatively passive," expectation was "decidedly an active state of consciousness." Memory sat with the past on a shelf, waiting to be recalled. Expectation, on the other hand, helped us meet the impending future. "All possibility of progress, physical, moral and scientific, depends on the power to anticipate," she wrote.

All possibility of progress. Hitchcock was an absolutist: Without expectation, she argued, people were utterly incapable of change—or of noticing it in the first place. We need the foothold. The climate can't change if you don't know what it felt like to begin with.

Yale had only begun admitting women to its graduate school in

1892, four years before the year of detection, and Hitchcock was one of the first three women to study philosophy there at the doctoral level. (Despite publishing as a psychologist, she conducted her work with the philosophy faculty, where her field was understood as a kind of "mental philosophy.") Today, it is easy to forget how revolutionary this moment must have been—how rigid were the expectations she bent. Men still wore top hats, and their idiotic starched collars had reached a towering high-water mark (three inches!). At the turn of the century, Hitchcock was possibility of progress incarnate.

But possibility is not the same thing as progress itself. All three women who earned doctorates in philosophy from Yale by 1900 would go on to take up teaching and mentorship roles—far less prestigious in academia than the research jobs found at elite universities like the one they'd just attended. None of the three would publish research past their dissertations; expectations would bend but not break. In the academy, Clara Hitchcock's groundbreaking work wouldn't be seen as such. Psychology was still coming into its own as a field, and the current trend was toward empiricism—toward tests. Her dissertation didn't contain enough experiments and would largely go unread.

Arrhenius, Kincer, Hitchcock. Names on a shelf waiting to be recalled. It's a small wonder people manage to accumulate any shared knowledge at all. It's much easier to forget—in this case, to forget how memory and expectation work in concert to link the past to the future; to forget that anticipation is a map, that climate is culture; to forget that in the absence of culture, generations are just trains passing in the long night of history. Five months before Joseph Kincer published his first essay on climate expectation, Clara Hitchcock died of breast cancer. I like to imagine the conversations they might have had. Instead, we have their neglected essays.

There's very little in neuroscience textbooks about the biology of forgetting. When it's covered at all, it's often couched in terms of the natural strides of aging or the aberrant fingerprints of neurodegenerative diseases like Alzheimer's. Canonically, forgetting is the passive degradation of memory, the fading of frescoes under the harsh light of time.

It isn't true.

In 2012, the Scripps Research neuroscientist Ron Davis noticed something peculiar in his laboratory fruit flies. Using electric shocks like Pavlov had used bells, he could train the flies to avoid certain odors. This much was a standard illustration of learning and memory. But he found he could also get them to permanently *forget* the associations in question by activating a specific subset of dopamine-producing neurons in their brains. It was as if the cells were sending a "forget" signal to the areas in which the memories were encoded. Conversely, blocking the release of dopamine from these cells ensured the memories stayed put: The flies continued to avoid the smells they'd learned to associate with shocks. Davis and his colleagues had taken an important step toward proving that forgetting was an active biological pursuit—as opposed to merely passive degradation—and that it was mediated by a particular group of neurons. In fruit flies, anyway.

Their results launched a thousand ships. Over the following decade, researchers would identify hallmarks of active forgetting in rats and mice; eventually, they would find them in people too. Mammals were more complicated than fruit flies, but Davis's findings appeared to be remarkably stable across species. Neuroscientists in Edinburgh, for example, showed their colleagues how they could stop rats from forgetting the locations of specific objects. Their experiment amounted to a biochemical trick—an intervention in a natural neural process. Here: Zoom in to a rat brain. To form a memory, it needs to form an association. Physically, this process often entails strengthening the connections between specific brain cells. To do so, neurons recruit proteins that form microscopic tunnels through which signaling molecules can flow into cells. What the Edinburgh scientists demonstrated was that if you chemically prevented the brain's active removal of these proteins, the connections encoding the rats' memories remained robust. They didn't forget.

The work was a reminder that memories are written physically in the brain. They are not amorphous puffs of spirit: They are constellations of neurons—neurons encoding specific sights and smells and words and emotions—whose connections have been strengthened. These constellations are called engrams. And if you can protect an engram, you can remember the story it tells.

A theory was beginning to take shape. Forgetting could scarcely be understood as mere degradation anymore. Instead, it was a competitive process—one that sparred with the forging of memory.

"Every species that has a memory forgets," reflected the cognitive neuroscientist Michael Anderson in a recent interview for *Nature*. "Full stop, without exception. It doesn't matter how simple the organism is: if they can acquire lessons of experience, the lessons can be lost. In light of that, I find it absolutely stunning that neurobiology has treated forgetting as an afterthought." Like Davis of fruit-fly fame, Anderson is a card-carrying member of the new cadre of neuroscientists working to ensure we don't continue to treat it as such. Just as Clara Hitchcock elevated expectation to the vaunted psychological peaks of memory a century ago, so, too, are the new mnemographers lifting forgetting to the same heights.

But the field is only a decade old, a comma in the history of memory studies, which for all intents and purposes began with Plato. And while its proponents have continued to amass reams of evidence in favor of the active-forgetting theory, a foundational question has begun to groan under the weight. Namely: *Why do we forget at all?*

Certainly, memory itself carries adaptive, evolutionary value. It doesn't take a neuroscience PhD to deduce that critters with memories—of foes, of friends, of food caches—are more likely to survive in the wild. It's less obvious why these same creatures might have an evolutionary need to forget. (You don't often come across someone who wishes they had a more fallible memory.) And yet the active-forgetting theory implicitly suggests there's adaptive value in a biological process that competes with the ability to remember.

Paul Frankland submits that we're wrong to consider memory and forgetting as wholly distinct processes. A neuroscientist at the Hospital for Sick Children in Toronto, Frankland argues that the two phenomena are twin components of a more general cognitive process of updating beliefs—and that it's the balance between these components that affords the most advantageous worldview. From this perspective, forgetting is a form of learning too.

Imagine never being able to forget. Memory without forgetting would imply a brain unable to generalize. It could neither extinguish faulty information nor respond to a changing environment. Writers have known as much for decades: Consider Funes from the Jorge Luis Borges story of the same name, the character locked in a mind of near-infinite resolution, "the solitary and lucid spectator of a multiform world which was instantaneously and almost intolerably exact." Funes's curse was to remember everything. It was immobilizing. He was bedbound. In Funes, Borges prefigured Paul Frankland's argument by eighty years: "Without effort, he had learned English, French,

Portuguese, Latin. I suspect, nevertheless, that he was not very capable of thought. To think is to forget a difference, to generalize, to abstract. In the overly replete world of Funes there were nothing but details, almost contiguous details."

The key is in equilibrium, suggest Frankland and Borges. Remember too much, and the hot-air balloon of our minds would never get off the ground. Forget too much, and we'd drift away into the clouds. It is only in delicate balance that we tune our knowledge to match the structure of the environment. In balance, we can float and steer. Forgetting is a form of neuroplasticity: Our brains are actively working to update our beliefs so we can best navigate the world as it changes. It's not just true of flies and rats. Neuroscientists studying people who have lived through earthquakes, for example, have noted that survivors often acquire better navigational memory after the fact. Their expectations of stability have been reconfigured. Yours would be too if you didn't believe in the ground anymore.

Frankland wrote with a colleague recently that the *rate* of forgetting depends in part on the degree to which the environment is predictable; that is, "in static environments, forgetting may occur less frequently as information remains useful, whereas in dynamic environments, learned information becomes less informative over time." If the point of active forgetting is to help us accurately model the world, then environmental change ought to trigger updates to the specific beliefs that are in conflict with reality. Brains target inaccuracies for suppression. "Indeed," they wrote, "not all memories are forgotten equally."

It's around this point in the science that I imagine a geographer like Mike Hulme getting a little sweaty. Hulme's "unsettling" of climate expectations is the neuroscientists' active forgetting. The same year the Edinburgh researchers were able to block forgetting in rats,

Hulme wrote that climate change "disorients our memories, many of which are profoundly attached to past weather, and unsettles our expectations about the future." In the language of neuroscience, we might say, when our expectations about the environment are violated, the errant predictions trigger forgetting plasticity, thereby modifying the accessibility of extant engram cells. In English: Climate change causes amnesia.

Applied to the idea of climate as expectation, findings from researchers like Frankland and Davis suggest that the psychological core of the climate problem is in its dynamism. Weather stories—shared weather memories—help us know what a place is about. As these stories bear less and less resemblance to place as we experience it, they become less useful for acclimating. And so our noggins adapt accordingly. Climate change is a discrediting force, whispering that the world has always felt this way.

Try it yourself.

How long is summer? Three months?

That wasn't true until around the turn of the new millennium. In the 1950s, climatologically speaking, summer in the Northern Hemisphere was fewer than eighty days long. The thing we call summer has lengthened by about four days every ten years you've been alive. Without significantly curbing emissions, climate scientists estimate it will last almost half the year by 2100. Winter will run less than two months.

A year after the year of detection, in July 1897, the SS *Portland* docked in Seattle carrying sixty-eight miners and two tons of yellow metal. The *Seattle Post-Intelligencer*, which chartered a tugboat to meet the steamer and break the news, would print a delirious headline: just four instances of the word "gold" in all caps, separated by exclamation points. The ship's arrival sparked a stampede to the Yukon; fortunes were made by Seattle outfitters who piled flour and bacon and clothing ten feet high in front of their shops. Of the hundred thousand prospectors who rushed north from Seattle to the boomtown of Dawson City, only a third would make it. By the time they arrived, many of the claims had been staked by people who already lived in the region.

Expectations come from our own observations of the environment—the way we feel in it. But they also come from what we're *told* about the environment: the stuff the headlines print, the happy people on the boats. In climate science, our SS *Portland*—our signpost for the way the world is supposed to work—is known as the "climate normal." This phrase refers to the statistical measurement that defines "average weather" in the first place. To understand the climate as changing, scientists must have something against which to compare it. Climate normals are thirty-year averages of meteorological observations like temperature and precipitation, corrected for missing data and faulty readings from the occasional misbehaving weather station. Mathematically, they define what constitutes the idea of the climate for a given region. They set our expectations.

They also modify our expectations. Since the advent of climate normals by the World Meteorological Organization in the early twentieth century (when it was known as the International Meteorological Organization), agencies tasked with computing the normals have regularly updated *which* thirty-year windows they use to calculate the averages. In fact, they're mandated to do so every ten years. This really happens. The thirty-year average is a moving average.

In a warming world, some climate scientists suggest that the widely accepted practice of continually updating the normals corresponds to moving the goalposts. As National Oceanic and Atmospheric Administration (NOAA) researchers wrote in 2011, "The key problem is that climate normals are calculated retrospectively, but are often utilized prospectively." We allegedly use climate normals to understand the present and the future, but if the past on which they're based is always changing, that means we're constantly injecting fresh amnesia into the system. If we updated climate normals frequently enough, it would look like the climate was never changing. We'd always inhabit the new normal.

It's not the view everyone takes, which helps explain the prevalence of the practice. "We want to give the best estimate for today's climate," argued the NOAA scientist Michael Palecki upon the release of the 1991–2020 US climate normals in 2021. "This is the dual nature of climate normals, being a ruler to measure against and a seer to predict the near future." Climate normals help us understand the environment as it exists today. Each new baseline "allows travelers to pack the right clothes, farmers to plant the best crops and electrical utilities to predict seasonal energy usage," writes Palecki. If farmers made their sowing decisions based on the 1931–60 normals, they'd be using the dust bowl as a model of what to expect in the early twenty-first century. (Which, unfortunately, may wind up being all too accurate.)

In the United States, the newest climate normals tell us that the country is warmer than it was a decade ago; that the eastern two-thirds of the contiguous states are wetter. But even a proponent of the updating process like Palecki notes that given the overlap of normals from one decade to another, "the changes are perhaps more muted than expected." It's only in taking a longer view that the warming is most apparent. Comparing the new temperature normals to the twentieth-century average, for example, reveals the jarring changes that demand our collective action, the changes that Joseph Kincer caught the first whiff of back in the 1930s and was so hesitant to accept.

Kincer's hesitance wasn't the fault of loose climate normals, though. It couldn't have been; they'd only just been invented when he penned his first essay on climate change. No, his doubt was feral—an animal burrowed deep in the cave of his body. It was doubt rooted in experience.

And doubting the world can be a beautiful thing. It can mean trusting that hidden animal inside you. Most heterodox voices and radical ideas come from confidence in a gnawing doubt, from deference to the self instead of the zeitgeist. But trouble arises if you aren't open to the animals of others—or to generational memory.

In 1995, a marine biologist named Daniel Pauly wrote a letter to the journal *Trends in Ecology and Evolution*, which the journal published in full. Pauly was working as a fisheries scientist, and he was frustrated by his colleagues. Namely, he thought he worked with a bunch of amnesiacs—and that the customs of his chosen profession catered to a kind of generational memory loss. He called it shifting baseline syndrome:

> *Essentially, this syndrome has arisen because each generation of fisheries scientists accepts as a baseline the stock size and species composition that occurred at the beginning of their careers, and uses this to evaluate changes. When the next generation starts its career, the stocks have further declined, but it is the stocks at that time that serve as a new baseline.*

What results, he wrote, is "a gradual shift of the baseline, a gradual accommodation of the creeping disappearance"—in Pauly's case, the disappearance and shrinking of fish stocks. With this accommodation comes the inability to assess the true impact of overfishing or to create meaningful targets for conservation or rehabilitation. A passing glance at any historical commercial fishing photo will validate his claims. They don't make the tunas like they used to.

It is unlikely that Pauly knew he was describing a fundamental aspect of human psychology. Yet since his naming of the syndrome in 1995, shifting baselines have been illustrated everywhere from conservation policy to political science. The World Wildlife Fund's Living Planet Index, for example—which tracks thirty-two thousand species of mammals, birds, fish, reptiles, and amphibians—revealed in 2022 an average 70 percent decline in animal population sizes since 1970. This is a staggering, seemingly unignorable figure. Yet people routinely fail to correctly estimate bird, fish, and invertebrate population trends. Baselines are always shifting. And of course they are. Our expectations are forged in the fires of experience—and others' experiences pale in comparison to our own lived truth. Or, as Pauly wrote, "the big changes happened way back, but all that we have to recall them are anecdotes."

Daniel Pauly's theory of shifting baselines is all the more resonant in the face of a changing climate. Is there a better description of our response to global warming than "a gradual accommodation of the creeping disappearance"? Environmental change can unfurl so slowly, some of it generationally. And while it's true that shifting baseline syndrome is well embodied by the updating of climate normals—it's basically baked into the pie at that point—NOAA can't carry all the blame here. Climate amnesia lives within us. You don't need to move the goalposts; to form expectations, you just need to be born.

It is a dirty synergy: active forgetting, climate normals, shifting baselines. They work in concert to obliterate the past. There is some surreptitious neurobiology here—the dampening of synapses to most accurately encode the present—but there is also the sad, small fact that you've only been alive as long as you've been alive. It is this baseline that is hardest to unlearn. The curse of only being in one body is that we only ever have our own reference points. You never shivered in Granddad's winters, try as you might to listen to him. You don't know the gold ship is a ghost ship until you get to the Yukon.

The blue of the ice at Jökulsárlón is like a neon sign advertising climate change. Five hours east of Reykjavík, the lake is the product of the melting Breiðamerkurjökull glacier, which has lost some sixteen cubic miles of ice over the last century—equivalent in volume to twenty-five thousand copies of the Great Pyramid at Giza. Every day, icebergs calve from Breiðamerkurjökull and slosh into the lake. You can stand at the lagoon's edge and watch slabs of electric cerulean drift southward into the Atlantic. You can follow them to a black-sand beach called Diamond and bend down and hold thousand-year-old ice in your hand. It is a tremendously haunting siren. They filmed a couple of Bond movies here. It feels like forgetting. It feels like the landscape is forgetting something too.

Aiding my understanding of the glacial lake today is an interpretive sign about climate change. (I am heartened to learn that "climate change" in Icelandic is "loftslagsbreytingar," a word of five huddling syllables that could only have been invented in the cold.) The sign includes photos of the outlet glacier in 1935 and 2015. The difference is one of kind, between ice and rock.

The old photo is helpful as a record of what came before. It is the opposite of a climate normal. There is no averaging going on here, no updating or shifting of baselines. The caption tells me it was taken in July—the same month my friend and I are at Jökulsárlón—only eighty-six years earlier. Kincer had just begun to wonder if the climate really was changing. I can make out two figures in the photograph, and they are tiny against the monstrous glacial edge. It looks like Granddad's winters.

If active forgetting implies we can't always trust our memories of the environment—and the slippage is further codified in shifting baselines and climate normals—it would seem we're left wanting for a record like this one. It is difficult to argue with a photo. With phenomenology. The glacier was just over there, and now it isn't. We should replace every mile marker on the highway with a before-and-after picture.

Can we remember what we've forgotten? In his lab, poking and prodding his fruit flies, Ron Davis, the Scripps neuroscientist, says yes. In 2021, his group reminded the world that although brains routinely forget, forgetting is not the same thing as deletion. Memories may indeed be permanently forgotten, but sometimes they are merely "temporarily irretrievable," the group wrote—"resulting in transient forgetting."

The transient version of active forgetting is something like a tip-of-the-tongue state—*what in the name of bulging eggplants is that movie called with the rat chef?*—a reversible suppression of something you know. And in fruit flies, transient forgetting appears to be mediated by a single cell on each side of the brain. Davis and his colleagues, rooting around in an organ the size of a poppy seed, had found it: a neuron that could temporarily suppress the flies' memories—in this case, memories of those Pavlovian smells and shocks—but that failed to quash the long-term associations in question. By activating the cell, Davis could effectively turn the memories off, as one might flick off a radio.

In people, reversing climate amnesia is hardly a matter of recalling a movie's name. We have generational and geologic memories to recall, and as we've seen, our brains are working against us in that regard. As Mike Hulme has written, rising mobility and migration, too, can shorten and weaken climate memories "as societies fragment and lose inter-generational cohesion." Displacement implies a place only existing in memory. And then we forget.

But as Paul Frankland and a colleague wrote recently, "Forgetting appears to be an active process of neuroplasticity that does not necessarily lead to engram loss." Just as the Davis lab has illustrated, Frankland stresses that forgetting isn't necessarily permanent: "Forgetting is potentially reversible, has adaptive functions, is modulated by environmental flux and is triggered by mismatches between expectations and the environment." You think the world is a certain way, it turns out it isn't, and your brain adapts accordingly. You are learning. But that doesn't mean your memories of the way the world used to be are gone. They may just be less accessible.

There are crumbs of possibility here, because we do have a capable tool for fostering collective memory, for reckoning with present contingencies. It just usually isn't found in neuroscience labs. We tend to call this tool "history." It is not an enterprise known for its objectivity. But maybe objectivity isn't exactly what we're after.

Yes, our brains have been evolutionarily tuned to accurately (one might say objectively) model the environments around them. Yes, neuroscientists like Davis and Frankland have shown that there is adaptive value in forgetting the old world—in stabilizing your expectations in a dynamic environment. But evolution is a poor map for a world changing at the speed of the fracking boom. Also, objectivity is a lie? Everything is interpretive, all the way down. We need more than adaptation to the present to respond to global warming. We need the long view; to thread a line from the past to the future.

History can be an act of elegy, but maybe it can be an act of evolution too. Maybe one of the things that makes us human is the potential to transcend some of the biology here. Realizing this potential requires reaching past ourselves, though, both into the past and into the lives of others. In his original 1995 essay on shifting baseline syndrome, Daniel Pauly offers a full-throated defense of anecdote—and of traditional knowledge. "For example," he writes, "astronomy has a framework that uses ancient observations (including Sumerian and Chinese records that are thousands of years old) of sunspots, comets, supernovae or other phenomena that were recorded by ancient cultures, and this has made possible the testing of pertinent hypotheses." For Pauly, astronomy was a sounder science because it took seriously the experiences of others, even if those others lived in the distant past.

That's what empathy is, isn't it? As Leslie Jamison notes in her essay on the subject, "Empathy comes from the Greek *empatheia—em* ('into') and *pathos* ('feeling')—a penetration, a kind of travel." History, empathy: They are modes of projection. They extend the brain's window of perception past its own frame. "Empathy requires knowing you know nothing," writes Jamison. "Empathy means acknowledging a horizon of context that extends perpetually beyond what you can see."

What we can see is often static. Climate empathy, the antidote to climate amnesia, requires a dynamism that matches not just the structure of our own environments, but those of others and of environments past as well. Pauly's argument for anecdote, for history, is a suggestion that much of the work toward this empathy will be social and communicative in nature—and will have more to do with memorial and storytelling than with the science of memory or the rejiggering of climate-normal formulae.

Some are trying. In England, geographers Georgina Endfield and

Simon Naylor have sought to make weather memories more permanent. The Weather Memory Bank is a digital experiment in cataloguing British experiences of the weather and their perceptions of climate change: an internet platform stuffed with video interviews and discussion questions to spur reflection on the other side of the screen. (It is fitting that the project has its home in Britain, where the psychologist Trevor Harley has suggested the casual interest in talking about the weather "borders on an obsession.")

In the video I'm watching, Derek and Vic, the two interviewees, perhaps in their seventies, are in rain jackets and standing in front of what appears to be a corrugated steel door of some flavor. It's unclear which is which. The interviewer is behind the camera, firing questions at the Brits. It's all a bit disembodied.

> **INTERVIEWER:** *How would you describe the weather where you live?*
>
> **DEREK** (maybe): *Rain, cold.*
>
> **VIC** (maybe): *Wet.*
>
> **D:** *Horrible.*
>
> **V:** *Windy.*
>
> **D:** *Yeah.*
>
> **V:** *At this moment.*
>
> **I:** *And more generally?*
>
> **V:** *It's changed. A lot. It's changed a lot. We're into, what, May now, June, and I can't get in the garden because the grass is always wet so you can't do anything. Same with you, isn't it? . . . You can't go on holiday anymore without taking an umbrella and some heavy clothes.*

43

It's not exactly *Don't Look Up*. I have no doubt that Vic (or maybe it's Derek) is telling the truth; that his experience of the weather has changed; that his umbrella collection has ballooned. But I'm not really sure what to do with this information. Maybe a future generation will. Sitting in front of my laptop in Seattle, clicking through interviews with Brits who volunteered for the Weather Memory Bank in 2014, I'm left wanting a little more. So much of living under climate change is wanting a little more.

Historians who have considered their profession's potential contribution to climate empathy have focused on the use of parable. Those who study the past can offer the rest of us parables of collapse, for example—stories of ancient societies caving to the whims of nature, to droughts and floods and crop failures; suggestions of what may come to pass if our own society fails to make the requisite changes. But as the historian David Glassberg notes, a spotlight on collapse risks a kind of paralysis—a lack of action rooted in a gloomy determinism. Why push for change if the future is already written? So, Glassberg writes, historians can also communicate parables of sustainability: tellings of more harmonious relationships between our species and the rest of the world.

But he is unsatisfied here too. "Parables of sustainability offer the public hope that the harmony with nature allegedly enjoyed by 'traditional' societies can be rediscovered," he writes, "though the nostalgia for an idealized past underlying these stories often turns them into tales of rupture, disconnected from the reality of the present situation." Idealizing indigeneity and land reciprocity without a serious mechanism for redress—that's just homesickness (emphasis on the "sickness").

Instead, David Glassberg argues for parables of resilience. History is brimming with examples of peoples rebounding from natural disasters and disease and war. And these narratives are different from those of collapse and sustainability, in that "they neither romanticize the past nor imply that it is too late to avoid a pre-determined dystopian future." They illustrate a human capacity to adapt and rebuild. They remind us that change is possible and that it doesn't need to rely on saviorship or fantasy.

In Iceland, my friend and I hadn't quite been able to make it to Okjökull, the dead glacier northeast of the capital. Unlike Breiðamerkur-jökull in the south, Okjökull had been small to begin with. In 2014, it had lost so much mass that glaciologists determined it no longer met the definitional criteria of a glacier. It was rocks interspersed with occasional ice, and the ice wasn't thick enough. They deregistered the glacier seven hundred years after it had become one.

When activists caught wind of the change in status a few years later, they produced a short documentary film about the life and death of Okjökull. They interviewed Icelanders about the landscape, about what glacial death meant to them. And then they held a funeral up on the old shield volcano where the glacier had once reigned.

Children present that day at Okjökull pressed a bronze memorializing plaque onto a slab of basalt at the site of the former glacier. It was a makeshift tombstone standing in protest of forgetting. A different kind of weather memory bank. The funeral implied there were others worth saving, that death doesn't always need to imply loss. As the Icelandic writer Andri Snær Magnason observed in interviews for the documentary, one of the most striking aspects of contemporary life is that geologic time is merging with human time—that we can watch stories that took centuries to write now rewind in a generation.

Can we rewrite a glacier? How do you compose a history of forgetting? The plaque is adorned with Magnason's words, first in Icelandic and then in English:

A letter to the future

Ok is the first Icelandic glacier to lose its status as a glacier.
In the next 200 years all our glaciers are expected to follow the same path.
This monument is to acknowledge that we know
what is happening and what needs to be done.
Only you know if we did it.

Ágúst 2019
415ppm CO_2

2

WET MACHINES

We may look into that window on the mind as through a glass darkly, but what we are beginning to discern there looks very much like a reflection of the world.
    —Roger Shepard, *Mind Sights*, 1990

The earth and myself are of one mind. The measure of the land and the measure of our bodies are the same.
    —Chief Joseph (Niimíipuu), statement to the US government, 1877

don't realize I'm in a graveyard until I'm stepping on bones. It's a beach on the Kitsap Peninsula, in Washington, and what's usually indigo has gone alabaster: miles of sand dollars, drained of their deep purple and piled atop one another like there'd been some kind of stampede at the exit. Rock crabs, too, slow-cooked in their shells; jellyfish, boiled beyond recognition and plastered to the sand. It was a bloodbath that smelled like a clambake. Reports on the late June

2021 heat wave had said a full billion marine animals had died on the coast that week. It had been the same temperature in Seattle—107 degrees Fahrenheit (42°C)—as it had been in Phoenix. Two hundred miles north, in British Columbia, the mercury hit 121 degrees Fahrenheit (49°C), and when a spark found its way to the tinderbox that was the town of Lytton, 90 percent of it burned. On the Kitsap beach, every step is a shattering.

I'm walking with Dianne Croal, a landscape architect recently retired from the National Park Service, who's telling me about the time she saw a tree explode over a river. Spontaneous combustion: In a heat wave, the pine trees are hair triggers. Terpenes beneath the bark harbor enormous potential energy when heated, and at sunset, when the light glints off a river at just the right angle, like through some kind of absurd magnifying glass, this energy can be released in the form of a chemical hand grenade. It's the same reason why turpentine is so flammable: It's made from pine resin. "It was one of the scariest things I've ever heard," says Croal, pausing before making a deep, guttural whooshing and whumping sound, her eyes bugging out behind her glasses. It was the sound of detonating.

I look down at the crab shell I've just stepped on and wonder when, or if, I'd notice I were baking alive. Would it be slow and steady—a crab without shade—or would I ignite, pine-like, some effigy to short-termism? Applied to climate change, there is a small irony squirreled away in the fact that the colloquial phrase for shifting baseline syndrome is "boiling frog syndrome." I think of Norman Mackworth and his telegraph operators. Mackworth had made something of a name for himself boiling frogs.

It went like this: On a morning in 1945 or 1946—the exact date is lost to the sands of academia—five Royal Air Force personnel found themselves shirtless in a Cambridge laboratory. The men were trained in telegraphy, and most had just returned from three years in the Middle East theater and the Indian Ocean. The stakes were lower in Cambridge than they had been in Aden and Ceylon, but the job was the same: Seated next to one another at a cluster of austere wooden tables, headphones pulled over their ears and pencils in hand, the operators were transcribing the dits and dahs of Morse code. Tones streamed into the young men's headphones; they scribbled out what they heard: *J49Y4, B8C9M, 126T6, B7LUB*. The nonsense words were called SYKO, and each complete message consisted of 250 of them. At twenty-two words per minute, the rate at which the perforated paper tape encoding the SYKO was fed through a transmitter sitting in the hallway, the operators could get through nine of these messages every three hours. Most of the men tied sweat rags around their foreheads. The dry bulb thermometer in the room read 105 degrees Fahrenheit (41°C).

It was hot by design. The war had been hot. For the operators stuffed into the Cambridge lab, working temperatures abroad had regularly exceeded 80 degrees Fahrenheit (27°C). Royal Navy service members working below deck had it worse: British ships generally weren't air-conditioned, and temperatures on board could reach 100 degrees Fahrenheit (38°C) or more. Compounding the heat was the fact that war making had become a technical endeavor. There were new cockpit gauges to inspect, radar screens to monitor, signals to transcribe, codes to break. And soldiers were making mistakes. As Royal Air Force squadrons flew for miles and hours overwater, as ultramarine expanses washed by and the dull teeth of fatigue began to dig in, pilots would falter, missing the radar blips signaling the presence of U-boats below. Britain's military leaders began to wonder what they were losing when they forced their airmen to stare at their screens for hours on end; when they packed their operators into compartments and forced them to transcribe in a hothouse. They realized they knew very little of the psychological demands they were placing on their service members. And so the British brass called on a young man named Norman.

A budding Cambridge psychologist studying vigilance, Mackworth answered the call to service, intrigued by the question of how the environment acted on fatigue and the intellect. Soldiers became subjects. By controlling their tasks and the world around them—and tweaking environmental variables one by one—Mackworth could isolate external effects on cognitive performance. One of his first studies was the Morse code experiment. Baking young soldiers in his lab, he found hotter temperatures melted years of training away: a simple and stark diagram of heat's effect on cognition. The men were otherwise healthy, but it was like the heat was slowly sapping IQ points out of their ears. He turned up the heat, tick by tick. By just 92 degrees Fahrenheit (33°C), some of the operators were making so many transcription errors that Mackworth didn't have room on his summary chart to trace a curve illustrating their performance. He merely drew an arrow pointing off the page, toward oblivion. Crabs in their shells; frogs in a pot.

There's no evidence that Norman Mackworth knew Joseph Kincer, the interwar US Weather Bureau chief who'd written of the changing climate—or that he'd ever known Clara Hitchcock, the turn-of-the-century Yale PhD who'd researched the psychology of expectation. But we can weave a tiny thread through their work: a faint line connecting a changing environment to a changing mind. Hitchcock believed there was no life without active updates to one's worldview. Kincer offered us evidence that the world worth viewing was in a state of climatic flux. Would a mind respond in kind? Norman Mackworth had unwittingly begun poking at this question, and a tentative answer rung out of his wartime lab: *Yes, and it wasn't pretty.*

I'd last thought of Mackworth the Thanksgiving before the pandemic, when I'd called Sophie Date, a public school teacher in Philadelphia. She'd been driving to New York for the holiday. Merging, she told me, "I can assure you, teaching without an air conditioner is one of the hardest things to do as a teacher." I suspect Norman would have empathized.

She described her previous year. For teachers like Date in the city's main public school district, the first days of the 2018 school year were difficult. That August and September, the Northeast had experienced a heat wave of almost unprecedented magnitude. National Weather Service meteorologist Dean Iovino referred to the oceans as "bath water." By September 4, with the heat index tipping into triple digits, the Philadelphia School District had dismissed class early five times because of the greenhouse effect at work in its classrooms. "They couldn't ethically keep teachers and students in a building that's over one hundred degrees," said Date. "I think we had more heat days than snow days that year." Most of the buildings in the district are old; a little over a quarter had central air-conditioning at the time. Philadelphia councilwoman Helen Gym called the conditions "inhumane."

The inhumanity doesn't always warrant early dismissal. On days like these, when Sophie turns to face her students and the ghoulish heat together, her world history classroom feels less like an incubator of ideas and more like an incinerator. For safety reasons, the windows only open a foot, and so despite the light industrial drone of the classroom's two spinning box fans, temperatures in the room hover near 100 degrees Fahrenheit (38°C). Students adhere to a dress code that forbids shorts and open-toed shoes.

"Pretty much immediately, everyone is sweating," she said. "Their heads are down. Everyone is complaining that they're tired." Sophie is tired too, but because it is her job and there are ancient civilizations to tell her students about, she will take her place at the front of the room. "Then you go about your lesson. You sweat through it." Eyeballs glazing over to the tombs of the pharaohs; the ditches and slants of cuneiform; the Antikythera mechanism; Pizzaro's execution of Atahuallpa. Sophie will wear a linen dress, or dark colors, and every night upon returning home, she will shower and wash her clothes because she will have sweat through them each period.

Sophie Date is one public school teacher of about 3.1 million in the United States. Between public and private schools, she and her colleagues educate something on the order of 55 million students every year, each one plucked from those grand probabilistic distributions of wealth and power and trauma that together define what we call advantage and its opposite. The Philadelphia district in which Date teaches employs seven librarians for a population of 150,000 students. Many of the water fountains in her school are plastered with signs discouraging kids from drinking at their taps because their pipes were forged from lead.

The pipes are significant. Getting ahead in the country does not uniformly require the absence of lead plumbing, but clean water and its lack of brain-shrinking neurotoxins help. It also helps to be white, to live in an affluent neighborhood, to have access to public transportation, and to have been born to parents with college degrees. But the slow leech of lead poisoning offers a stark, instructive case study in how the physical environment of a student can bear on her education and economic mobility. The logic is artless and punishing: If you are unfortunate enough to grow up in an area in which the pipes and the paint are old and venomous, if in your human necessity you drink this water or in your human curiosity you eat these chips, then you have traded your full human capacity to learn. There is no safe level of lead in children. Microscopic amounts breed smaller chunks of cortex, lower IQs, and attention deficit hyperactivity disorder. In the United States today, we have consigned at least 1 million kids to a lead-poisoned fate of neurological damage.

"That's what working in an impoverished school system looks like," said Date. "We are asking people to learn and to teach, but we are asking them to do so in conditions that are not conducive to the work." She's talking about chronic underinvestment in the country's public school system, but she's also talking about organizing against the chaos of nature. "It's like if someone were to say you have to build this house, but we're not going to give you any wood or any nails," said Date. "Here's a hammer."

The lead pipe, the hammer: They make sense as weapons because you can curl your fingers around them. But climate change is jarring and nonsensical—self-immolating trees, beaches with jellyfish where the sand is supposed to go—and the bludgeoning of a warming world doesn't come from some underhanded Colonel Mustard type. Instead, the violence is wrought by the study itself. Since the 1950s, we've known that if you heat up a goldfish, or a nematode, or a mouse or a rat or an earthworm or a zebra finch, its long-term memory will suffer. Sophie's classroom, Mackworth's lab: They are visceral reminders of our animal selves. As temperatures rise, opportunity falls. Our brains don't work as well when it's hot out.

There is a measurement challenge," Joshua Graff Zivin told me. "And that doesn't make it any less real, though it does make it hard to tell your story." An environmental economist at the University of California San Diego, he speaks of measurement challenges as a mountaineer speaks of a new peak: There will be something good and pure up there at the top, but much of the thrill is in the route planning and the preparation and the hunt.

Graff Zivin, who has the wired look of someone you suspect was once struck by lightning, spent the early days of grad school probing the most cost-effective manners to design and meet various environmental standards. It was the mid-nineties in Berkeley—Green Day and the Offspring were taking Gilman Street punk to the top of the charts—and Graff Zivin's life was defined by weeks of brute-force number crunching: the type of calculus he can only describe today as tedious. As his mind began to wander, he started to wonder how we go about deciding what counts as poison in the first place. Too much chocolate will hurt you, but the EPA didn't regulate cacao nibs the same way it regulated DDT. The question led to a crash course in toxicology. He learned about chemical exposures and dose-response curves: the empirical relationships between how much of a substance you consume and whether or not it kills you. He learned the difference between toxins and toxicants. (It's a bit of a not-all-rectangles-are-squares situation.)

Armed with his new language, he translated the instruction manual for environmental regulation, where he read a naked truth. Environmental standards were strikingly blunt objects, devoid of nuance. They responded to the visible and the obvious: dead people and heart attacks.

He became obsessed with the notion that dramatic health outcomes were driving environmental policy—but that by the late 1990s in the United States, when he was in grad school, these outcomes were relatively rare. "It's not that anyone believes those are the only outcomes," Graff Zivin told me. "It's just that they're outcomes people can measure." Like the British brass surveying its troops at the height of the war, he began to wonder what was missing. And because he didn't have a Norman Mackworth to turn to, he started studying blueberry farms.

Imagine a fruit farm employing one hundred pickers. If two of these people are exposed to an insecticide and get sick, they won't come in to work the next day, and all else equal, the economic output of the farm will fall by 2 percent. It's in everyone's interest to label the insecticide a poison (not least that of the people sick in bed). Graff Zivin didn't disagree, but his hunch was that low-level, invisible effects of subtler environmental stressors like air pollution and heat, when summed across whole populations, added up to larger aggregate effects on health and productivity than obvious killers like DDT. As the Swiss physician Paracelsus had coined in the German Renaissance, "Sola dosis facit venenum." The dose makes the poison.

To get a sense of Graff Zivin and the father of toxicology's logic, imagine instead that the same farmworkers experience a workday in which ozone levels creep up by something like ten parts per billion—the kind of difference that's more or less undetectable without technical instrumentation. On average, if by some biochemical effect, everyone's output drops by just 3 percent, the economic impact of ozone on productivity would exceed that of the insecticide's, even if everyone still shows up for work. The workforce would be just as large, the downtick in output would be written off as random day-to-day fluctuation, and nobody would think to measure something as arcane as ozone levels. And yet some of Graff Zivin's early research demonstrated the reliability of exactly such an effect: In a study of California's Central Valley blueberry and grape pickers, he and a colleague showed that when ground-level ozone levels rose by ten parts per billion, productivity fell by more than 5 percent. People still showed up to the farm; they just weren't as good at their jobs.

As he and his peers honed a new field of inquiry, Graff Zivin bored deeper into the invisible. He needed to, because he was still working in the realm of the slightly obvious. Ozone forms when nitrogen oxides—found, for example, in vehicle exhaust and synthetic fertilizers—interact with volatile organic compounds produced through the use of fossil fuels. The ozone molecule irritates our airways and causes our lungs to glitch. Farming is hard work done outdoors. You don't need an intellectual leap to imagine that this combination would make someone less productive. But in the United States, on average, adults spend something on the order of 90 percent of their time indoors. More than 100 million Americans work in the service sector; another 20 million work in government. Graff Zivin suspected his environmental productivity effects followed people inside.

Evidence emerged like a gnat cloud. Air pollution slowed down everyone from pear packers to call-center workers. Higher temperatures tanked students' math scores, even when correcting for location and family wealth and, importantly, when comparing kids to themselves at different points in time (as opposed to one another)—one of the best statistical controls we have. During a heat wave in Boston, researchers showed that college students in dorms without air-conditioning performed cognitive tests with 13 percent slower reaction times than students living in air-conditioned buildings. There wasn't a toxin to speak of, just the poison of heat.

And where Norman Mackworth could only study five or ten people at once, the new environmental economists deployed their models on hundreds of thousands. Jisung Park, for example, an economist at UCLA, recently pieced together a dataset of 1 million New York City students—and calculated that higher temperatures on exam days had spurred at least a half million failed tests over a fifteen-year period.

Park had chosen New York because it represents a near ideal education laboratory: The population is enormous (New York's system is the largest in the country), there's ample diversity of experience across the school districts, and over the course of two weeks in June, every student takes mandatory exams at the same time. But despite this uniformity, the effects Park measured were far from uniform, not least since poorer public schools are less likely to be air-conditioned. According to his research, only two-thirds of New York City public schools had some kind of air-conditioning system at the time he conducted the study. The other schools, just like Sophie Date's in Philadelphia, are subject to the punishing whims of heat waves. Pointing to the country's disparate infrastructure investments, Park has estimated that the cognitive effects of hotter days account for roughly 5 percent of the racial achievement gap in the United States.

The phenomenon is global. In China, Graff Zivin and his colleagues assembled a sample of 14 million observations of students who took the National College Entrance Examination, or gaokao, which plays a high-stakes role similar to that of the SAT in the United States. Described as "almost the sole determinant for college admission in China," the test is administered in June and offered only once per year. In another echo of the new environmental-economic drumbeat, the researchers illustrated that a standard-deviation increase in temperature resulted in a 1 percent decrease in gaokao scores. A similar deviation also decreased the probability of acceptance to a top-tier university by 2 percent, leading the authors to conclude that "hotter regions may be unfairly penalized by the current system." Similar results have been shown in India. "Students are not any less intelligent on hot days," the researchers wrote. "They simply struggle to access that intelligence."

It's in this manner that heat is coming to be recognized as the subtle poison Graff Zivin suspected it had been all along. It is regressive and silent. It is cognitive death by a thousand cuts. And as Graff Zivin notes, it shares these characteristics with its epidemiological cousin: air pollution.

Park's work in New York and Graff Zivin's in China were inspired in part by an Israeli researcher named Avraham Ebenstein, who in a seminal paper in 2016 demonstrated that Israeli students' high school exit exam scores slumped with rising particulate-matter levels. In particular, Ebenstein's group showed that scores on the Bagrut—the Israeli analogue of the SAT and the gaokao—fell by nine-tenths of a point for every standard-deviation increase in particulate matter (relative to a day with average air quality). At the worst levels of exposure, the effect could offset that of decreasing class size from thirty-one to twenty-five students. Perhaps more worryingly, in an illustration of the gravity of the test and the seriousness of perturbations in one's score, Ebenstein also showed that increased particulate matter on testing days—but not other days of the school year—was associated with a lower probability of enrolling in college and lower wages later in life.

Following this example, UCLA's Park recently calculated that for the average student, testing on a 90-degree-Fahrenheit day decreases the likelihood of passing a given subject by about 10 percent. And failing any course tanks the likelihood of graduating on time. According to Park, a standard-deviation increase in temperature—about 4.5 degrees Fahrenheit in his sample—decreases one's chances of graduating on time by about 4.5 percent (compared to the average on-time graduation rate): a percentage point for every extra degree Fahrenheit at exam time.

That was a lot of numbers, so let's zoom out for a moment. These high school students—with college and careers and whole life trajectories in front of them—they represent one of the populations most vulnerable to the scrabbling claws of heat and air pollution. But the hunt is hardly limited to the classroom. "Basically every rock we turn over, there's something there," Solomon Hsiang, a UC Berkeley environmental economist and pioneering collaborator of Graff Zivin's, told me. "The climate just affects everything." It's hardly an exaggeration. Social scientists have shown environmental factors like heat and air pollution act on everything from sleeplessness to baseball umpires' error rates to investors' risk-taking behavior to the complexity of speeches made by Canadian members of Parliament. And in a temperature study of immigration judges, a research group at the University of Ottawa found something else disturbing.

In the United States, asylum decisions are made by a roster of just over 250 judges distributed scattershot across forty-three cities around the country. Despite these judges and their rulings technically being subject to the oversight of the attorney general, asylum decisions are generally understood to depend solely on the discretion of a given judge in the face of the facts of a given case.

The Ottawa researchers, led by the economist Anthony Heyes, sought to stress-test this dynamic: first by investigating whether these kinds of decisions, "the substance of which have nothing to do with contemporaneous temperature," might indeed depend on such an ostensibly irrelevant factor. He did what all good economists do: He amassed a dataset of hundreds of thousands of examples of the phenomenon under investigation. Stacking up two hundred thousand court decisions from over a four-year period—and isolating the causal influence of temperature alone—Heyes and his colleague ultimately found that for every 10-degree-Fahrenheit increase in outdoor temperature during a hearing, the probability of a judge handing down a decision favorable to the asylum applicant fell by almost 7 percent.

"To put this into perspective," they wrote, "in our sample, the difference in grant rate between a judge at the twenty-fifth percentile in terms of leniency, and one at the seventy-fifth percentile, is 7.9 percent." Temperature made a complete mockery of objectivity.

It wasn't a fluke. As Ebenstein had shown in Israel, Heyes's effects were limited to day-of exposure. It didn't matter how hot it was the next day, for example, or how hot it was in a given region on average; Heyes showed that only contemporaneous exposure provoked the effect. When I spoke with him via phone, he told me the results were galling in how arbitrary they seemed. He'd needed a manner of fact-checking his work. And so the Ottawa researcher repeated his efforts with a dataset of parole suitability hearings across California. Lo and behold, on hotter days, as with asylum decisions, the chance of a given commissioner granting parole to an incarcerated applicant lurched downward. The data had unmasked a dark capriciousness, even among some of the country's most ostensibly level-headed decision-makers. In their paper, Heyes and his colleague call the channel of influence "subtle and pernicious."

Subtle, pernicious, and everywhere. Back in the San Francisco coffee shop where we'd met, Berkeley's Hsiang asked me if I'd "seen the one about the chess players." I hadn't, but I could guess what it said. "People are pretty sensitive machines," he mused.

Sensitive machines. On the Kitsap beach, I pick up a porcelain sand dollar and am struck by its delicacy. I remind myself again that I'm looking at bones. Alive, *Dendraster excentricus* is covered in undulating purple spines, like a lavender farm in the wind. It has five rows of tube feet, sixty muscles, and five little articulating jaws that Aristotle thought looked like a lantern. (Today, all taxonomically related sea urchins' jaws are called Aristotle's lanterns for this reason.) Sand dollars eat algae. They reproduce by billowing out clouds of eggs or sperm; in a gentle marine dance, ocean currents complete the fertilization. Removed, a sand dollar's teeth look like birds in flight—so much so that in Christianity they're traditionally known as "doves of peace." The skeleton of a sand dollar, this thing in my hand, is referred to scientifically as a "test." As with students, all we tend to see is the test.

Beneath the test—the SAT score, the asylum decision, the chess move—is a sensitive machine. Cognitive attention, our ability to focus, doesn't just passively emerge from the mists of consciousness. It is a soaring, meticulous ballet that our brains have been rehearsing for hundreds of thousands of years.

Over the past three decades, neuroscientific efforts have broken the choreography of attention down into three main brain networks: a primitive system responsible for arousal and warning, a sensory system that directs our attention in space, and an executive control network that sustains our attention, helps us switch between tasks, and resolves cognitive conflict.

In line with their unique roles, each attentional system is more or less anatomically concentrated in a set of distinct brain regions. Alerting, the oldest of the networks, appears to be rooted in deep, subcortical areas like the thalamus, a relay center just above the brain stem. Orienting, on the other hand, is largely a function of sensory integration areas like the two superior parietal lobules, together a heart-shaped region on the top of the brain that combines visual and tactile inputs; as well as the frontal eye fields, which are responsible for the precision of visual tracking. Executive function, by and large, takes place in the front of our cognitive machinery: swathes of evolutionarily new cortex beneath the forehead, including the anterior cingulate cortex, known for its roles in error detection and conflict resolution. The cingulate is named as such because it wraps the neural superhighway linking the brain's two hemispheres; "cingulate" comes from the Latin word for "belt."

Cognitive neuroscientists have been able to localize these networks because they no longer have to wait for an opportune craniotomy to peer into the brain. A working organ needs oxygen, and because brain areas working harder than others in a given moment require comparatively more of the stuff, researchers can get a rough sense of which regions are exerting effort by tracking the main protein carrying the molecule: hemoglobin. A fundamental component of red blood cells (and the ingredient that dyes them red), hemoglobin becomes slightly less magnetic upon donating its oxygen to a given brain area. If you have access to a powerful enough magnet, you can tell the difference between oxygenated and deoxygenated hemoglobin by firing radio waves at someone's head and perturbing the proteins' magnetic fields, the wiggles of which you can then measure with your magnet. Stitch enough of these comparisons together over time and space, and you can trace blood flow—and, by extension, neural function—throughout the brain. Called functional magnetic resonance imaging (fMRI), the technique is regularly leveraged to dissect the brains of awake people without a scalpel ever being lifted. And in fMRI studies of extreme heat's effects on this neural hardware, neuroscientists have revealed temperature's ability to disrupt the third attentional network, governing executive function, but not the other two.

One of the things immigration judges, chess players, and students have in common is that they're engaged in relatively complicated cognitive undertakings. In a delicate, twisted paradox, what we're learning about the effects of heat on cognition is that they're most pronounced for complex tasks like arithmetic, motor coordination, and executive function—as opposed to simpler acts like basic memory recall or shape recognition. In the sea of thought, the first vapors to boil off are the most precious.

Nadia Gaoua, a sports-medicine researcher based in Qatar, has shown that people in her lab are worse at distinguishing between complex visual patterns or recalling specific spatial sequences when exposed to extreme heat—but can still do things like match simple shapes to one another or press the appropriate arrow key when prompted. In a complement to her research, Chinese neuroscientists have exposed volunteers to extreme heat and documented decreased blood flow to the anterior cingulate. This same group has also shown that heat reduces the connectivity between the brain regions involved in executive control. When people are exposed to high temperatures, the anterior cingulate unbuckles its activity from that of other cortical areas. The brain's fireworks become less coordinated and more randomized. Unbelted, thought sloughs off.

According to Gaoua, a modern-day Norman Mackworth and a gumshoe on the case of temperature-sensitive cognition, that's because heat acts as an additional cognitive load. In the juggling act of attention, a heat wave adds another bowling pin to the mix, and the pins become easier to fumble. Indeed, in her lab, Gaoua has shown that high temperatures boost an electrical measure of cognitive load in the brain.

So it's tempting to read her studies as evidence of a sad truth: that brains just aren't as good as they ought to be at adapting to higher temperatures. We reach a point at which we can no longer acclimate, and our cognitive performance suffers for it. But this reading isn't quite right, because it ignores the crown jewel the brain is working hard to protect: the biological miracle of its operation in the first place. Cerebral tissue is some of the most sensitive in the body to heat. At higher temperatures, brain cells struggle to metabolize glucose, the organ's primary source of energy. Heat can also overexcite the brain, leading to seizures, as well as cause misfolds of the tau protein, the tangling of which is one of the hallmarks of Alzheimer's disease. Around 102 degrees Fahrenheit (39°C), the structure of brain tissue changes. Cells begin to look abnormal. By the average July temperature in Qatar, some of them irreversibly stop working.

It takes energy to avoid these outcomes. Brain temperature regulation is far from static: It's an internal war against disorder, fought every moment of every day, with life-or-death stakes. While accounting for only 2 percent of the body's total mass, the brain burns a fifth of the body's glucose and consumes a quarter of its oxygen. All energy metabolized by the organ is ultimately released as heat: Every minute, the average brain produces about nine hundred joules of heat energy—enough to power a fifteen-watt lightbulb over the same time period. If you could capture the heat your brain releases over the course of a given week, you could warm enough water for a ten-minute shower. The nervous and circulatory systems need to work hard to keep the brain cool, and when it's hot out, they need to work harder. Perhaps there's a small paradox in our jettisoning some of our most prized cognitive developments upon exposure to just a few more degrees of heat—but only from the perspective of the outside looking in. On the inside, stability rules everything. And when your house is on fire, you take the essentials and run. It's not that the brain isn't working at higher temperatures. It's just working on a different problem: the more fundamental problem of trying to keep us alive.

Seen in this light, the discomfort we feel in the heat and our simultaneous cognitive decline are more than coincidence or imperfection. Discomfort is the blare of a psychological, evolution-arily tuned fire alarm. It's called an alliesthesial effect: Any environ-mental stimulus that pushes the body away from its equilibrium is perceived as unpleasant. The brain will do just about anything—including sacrifice hard-won hallmarks of humanity—to maintain its physiological status quo.

As I walk along the Washington beach, crunching my way over crab shells and dead sand dollars and thinking of exploding trees and Norman Mackworth, this fact—that our cognitive decline in the heat is a function of our brains fighting to save us—is one that glimmers overwater with a certain shade of optimism. The difference between crabs and people is that we don't have to rely on intergenera-tional evolution to survive as a species: We can plan and adapt within our own lifetimes. But doing so requires that we accept the problem as it is.

Here's how it is: We are lying to ourselves.

Our core lie is that people won't actually feel the strain of a cou-ple of extra degrees of warming; that the only thing acting on human behavior is the behavior of other humans. "In economics, people aren't trained to think about the physical world—they're trained to think about decision-making and the mind and incentives," said Sol-omon Hsiang. "There's this belief that if we get incentives right, any-thing is possible." It's research from people like Nadia Gaoua, Josh Graff Zivin, and Hsiang himself that scoffs in the face of this imag-ined rational actor.

Countering this lie requires a recognition of our porousness to the world. If you are the type of person who is reading this book, you are also perhaps the type of person who is used to thinking this way—or at least the type of person who is interested in doing so. If you'll permit a broad brush, though, Western societies tend not to be socialized to pay attention to these kinds of relationships. The gospel of individualism is rooted in a certain separateness from the world. But you are not an island.

Even in the corners of the academy where people study this kind of thing, researchers often frame their studies of heat and cognition in terms of productivity and economic output. In the parlance of economics, when you invest in climate adaptation, you're investing in "human capital." It's upon this fulcrum that the whole argument pivots, says Josh Graff Zivin: "The EPAs of the world can turn to the commerce departments of the world and say, 'Hey, this environmental regulation you thought was a stranglehold on growth is actually an investment in growth'—in the same way that building the interstate highway system was an investment in growth."

And that's all well and good, but I think we probably lose something sacred with this language: the intimacy and reciprocity of it all; that porousness; that effortless fluidity between environment and body and mind; that delicate manner in which the crackle of thought ricochets from sulcus to spinal cord to synapse to muscle to ink on the notebook page. To reduce learning to human capital accumulation is to minimize the magic of these leaps of faith. The fable of endless growth is what helped begin to boil the oceans in the first place.

No need to romanticize too much, though. I get it: It's a strategic argument. And in imagining the hundreds of millions of students of the world—at their desks, gripping pencils, squinting at chalkboards, maybe learning about gerunds or hypotenuses for the first time—even the sanitization of language does not hide the fact that education is an intense human effort and that some people aren't being afforded the opportunity to explore their full human potential. That much is worth remembering. Anthony Heyes, the Ottawa economist who led the research on immigration judges, writes about the possibilities of an emerging research agenda—ultimately, the melding of economics and biology—that "seeks to model the agents that populate economic textbooks as biological organisms ('wet machines'), sensitive to the environment in which they function." As we heat up and dry out, it is a wonder we ever imagined ourselves as anything else.

In an oft-quoted 1999 interview with *New Perspectives Quarterly*, when asked about the key to Singapore's success, the country's founding father and first prime minister, Lee Kuan Yew, called out air-conditioning:

> *Air conditioning was a most important invention for us, perhaps one of the signal inventions of history. It changed the nature of civilization by making development possible in the tropics. Without air conditioning you can work only in the cool early-morning hours or at dusk. The first thing I did upon becoming prime minister was to install air conditioners in buildings where the civil service worked. This was key to public efficiency.*

At Sophie Date's school in Philadelphia, an alumni class had the same idea as Lee, pooling their resources after the 2018 heat wave and outfitting every classroom with AC units. She tells me how environmental issues and building safety have become central issues for union mobilizing and contract negotiation, as well. But not every district has a strong teachers' union, and not every school has an alumni class bankrolling infrastructure investment. There's also more to the story of educational attainment and economic mobility than the climate. Lee's famous nod to air-conditioning was his second answer to the question. His first was a tribute to what he saw as Singapore's social contract of mutual, multicultural respect: "Otherwise, there can be no common progress. If you want to beat the other fellow down and insist that he act like you and observe your taboos, then the whole place will come apart. A live-and-let-live contract is thus a social precondition." In countries like the United States, the precondition isn't covered.

Besides, once we get past the first one, there's a second lie we tell ourselves: namely, that people are infinitely adaptable. Short-term responses to heat stress can help, but they will get us only so far. In Qatar, Nadia Gaoua has shown that strapping cold packs to people's foreheads can restore some of the cognitive capacity lost at higher temperatures. Amusing psychological evidence suggests that looking at images of cool environments is enough to moderately improve cognitive control during extreme heat. But these examples are Band-Aids. Researchers estimate that traditional modes of adaptation to global warming can rescue at most half of our climate-fueled health losses.

We also can't bank on evolution. By definition, it takes generations to, say, grow more sweat glands. Especially in the tropics, where humidity can make it impossible for sweat to evaporate, extreme heat will only become more dangerous in the coming years. In the world's worst-case emissions scenario, a third of the global population will live in areas as hot as the hottest parts of the Sahara within half a century. Our bodies can't adapt to that. As Hsiang put it, "There are only so many layers of clothing you can take off."

"Air-conditioners aren't going to solve all of it," said Sophie Date.

"Using a reusable shopping bag isn't going to solve it." There's also the small, twisted irony that expanded use of air-conditioning is likely to only increase peak electricity demand, require the construction of additional power plants, and—given the current global energy mix—exacerbate the use of fossil fuels. It's a textbook vicious circle. Even if we project ourselves into the utopia of a renewable-energy-powered future, air-conditioning still doesn't offer a silver bullet, because, as Anthony Heyes and others have shown, the effects of heat can follow us indoors. "How much of the time are you actually able to protect yourself?" asked Hsiang. "And how long does it last? If you go outside and work strenuously in the heat to run to class and you sit down to take a test, you probably don't reach thermal equilibrium before the end of it."

It's this dilemma that led me to the desert.

You wouldn't know Xero Studio used to be a dentist's office. The twenty-five-hundred-square-foot building on the outskirts of Phoenix, across the street from the baffling Z's Greek Chicago Style restaurant, has more in common with a spaceship or a shipping container than with a former medical facility. It's slatted, geometric, and mysteriously lit. Under its scrim of finlike vertical wooden planks, a perfectly temperate office stands in protest of the desert heat. Nine months of the year, skylights are choreographed with the studio's doors to offer natural, passive ventilation. In concert with the windows, they circulate cool air toward desks and meeting areas and release hot air through the roof. Careful studies of the sun's path ensured the studio could be naturally lit for most of the year without overheating. Outside, the wood fins cut down on glare and heat gain, and a tall desert hedge shades another wall of the building. Between the reuse of the old dentist office's masonry, new solar panels, an irrigation-free garden, and a water-recycling system, Xero Studio hits a triple net-zero threshold: net-zero energy, net-zero water, and net-zero waste. There's very little else like it in the world. The architects point to it as an example of regenerative, bioclimatic design—architecture that doesn't just mitigate climate damage but can help reverse it.

The studio is the professional home of Christiana Moss and the architecture firm she cofounded in 2003. The desert was the ideal place to launch a new practice, she tells me over the phone: "The natural environment is something you can't ignore here." Moss, who was raised in New York City, grew up viscerally aware of climate change— and of the design missteps we've made over the years that have dug the hole ever deeper. She recalls the subway stations of her childhood summers as almost improbably cool; it wasn't until recently that she realized air-conditioning was to blame for their modern mugginess. "We air-conditioned the subway cars and blow the hot air into the stations," she muses. "We keep making these mistakes over and over and over again." Not so with Xero Studio. It was her chance to put her values into practice.

For Moss, one of these values is "sustainability at any scale." Triple net-zero buildings shouldn't be reserved for the superrich—or for architecture firms. And by her estimates, they don't have to be. "This isn't rocket science," she says. "We just need to be really smart about simple things." Her guidelines for regenerative, climate-positive design are all about leveraging the natural world around us. You have to capture energy from the sun. You have to understand how heat and light flow through your site before building, and you have to design for natural convection. Open the windows at the right time of day (which is to say: at night). Plant deciduous trees and take advantage of their shade. In her words: "Work with nature instead of against it."

It's not exactly rocket science, but it's also not mysticism. Many of her firm's projects leverage thermal-imaging tools and daylight-analysis software to map a site's physical evolution over the course of a given day. With an understanding of these patterns in hand, the architects can calculate the optimal placement and orientation of windows so as to maximize daylight but minimize heat gain. "Outsulation" solutions like the wooden fins at Xero Studio help manage temperatures indoors; raised or sunken features on a building's facade can help it shade itself. Rocket science or not, as the world warms, Moss's efforts and insights in the desert will only become more applicable outside of it.

The challenge of building in extreme environments has always offered a lesson for the rest of us, she suggests. Traditional wind-catchers in the Middle East—open turrets that would drive cool air into a building—had been in use since at least the thirteenth century BC before their abandonment one hundred years ago in favor of nascent air-conditioning systems. Today, the ancient technology is having something of a renaissance. As temperatures rise—and the coronavirus pandemic highlights the importance of natural ventilation—windcatcher-inspired designs have cropped up from London to China to Malta. In the latter case, a local brewery saw a gap of 25 degrees Fahrenheit (–3°C) between peak outdoor temperatures and those inside its processing facility. Windcatcher-like designs can circulate air for twenty-five stories. Imagine if we redesigned schools with these goals in mind.

That's what Christiana Moss is trying to do, anyway. Today, she's a university campus architect. In her most recent project, she's taken on the Arizona desert climate again, working with Arizona State University researchers to design a framework for a triple-net-zero university laboratory connected to Tempe's transit grid. It's an opportunity to put her values into practice once more: The research facility will generate all its own electricity, and wastewater recycling and indoor gardens will produce potable water for the whole building. Her attitude on regenerative design is slightly vampiric here: "We don't want any direct sunlight to hit a piece of glass—ever." It'll be Xero Studio on steroids. The focus of the lab's future students? Climate change.

3

WHO KILLED
TYSON MORLOCK?

For now, these hot days, is the mad blood stirring.
—William Shakespeare, *Romeo and Juliet*, 1597

All I ask is that, in the midst of a murderous world, we agree to reflect on murder and to make a choice.
—Albert Camus, *Neither Victims nor Executioners*, 1946

In the waning days of June 2021, at the same moment hundreds of millions of animals were roasting alive on the Pacific Northwest coast, a man in Portland, Oregon, was hooking up an inflatable swimming pool to a fire hydrant. Tyson Morlock was from Washington but had moved a few weeks previously to the City of Roses, where he was living under the 99E overpass. It was hot in his new home. The heat wave had crested at 116 degrees Fahrenheit (47°C) in the city, and Morlock, no fool, had taken it upon himself to jury-rig a cooling site big enough for the whole camp under the expressway. If you were to have stood at the intersection of Southeast Sixth Avenue and Division Street that week, you would have heard the joy of splashes and the smack of water balloons and the unmistakable anarchy of at least one wet dog. For a few halcyon days, Morlock and his neighbors had a backyard pool.

I couldn't interview Tyson Morlock for this book because I only learned his name after he'd been stabbed to death in the early morning of July 1. It was a friend of his who did it, a man named Mark Corwin, the man with whom Morlock had moved to Portland. Apparently there had been some kind of dispute over a pack of cigarettes. In Corwin's telling, Morlock had flown into a rage upon being questioned about the missing cigarettes and had come after his friend with some metal tent poles. Corwin ran until he couldn't, and then he turned and planted his feet, a knife in his hand. He'd later claim self-defense; a grand jury would vote to acquit. "There's not a lot of places you can go in the community where you can feel safe," Morlock had told a journalist who'd interviewed him about the pool the day before he died.

It is tempting to pick at the scabs of causality here. Follow the knife into Morlock. Is it the punctured lung that kills him? Well, no, that doesn't seem quite right; the lung itself is cradled by a dense circulatory network of veins and arteries. Surely it is the loss of blood that brings him down. But that idea also feels like a little bit of a truism, right? There wouldn't have been any bloodshed without broken skin—without the knife. The blade, surely, is the cause. And yet . . . we know from the gun lobby that *only people kill people*, that knives don't act on their own, that there's some human agency here—hence our knee-jerk understanding of the death: Corwin killed Morlock. Clear enough. And even then, the causality feels hazy if we think about it for too long. Corwin had called emergency services immediately upon issuing the injury. He told the dispatcher he didn't know what to do. He was trying to stop the bleeding with his own T-shirt. It's hardly the picture of a murderer.

So was it really Corwin? The grand jury called his self-defense "justified." This designation further muddies the waters. In the jury's reading, it's not Corwin killing Morlock at all: It's Morlock killing Morlock. Corwin is just a vector for Morlock's rage: a kind of deadly ricochet. And maybe we could market this version of the story as intuitive and palatable too. We could point to Morlock's recent history of methamphetamine use; twist his homicide into a story of personal responsibility ready-made for Fox News. Switch to MSNBC and point at the Pacific Northwest's housing crisis. If Amazon hadn't set up shop in Seattle, if housing supply had kept pace with rising incomes and a population boom, maybe Morlock would have remained stably housed in Washington and never ended up sleeping under an overpass in Portland. Maybe he'd be alive today.

Heck, if missing cigarettes lit the fire, perhaps the fault is in their absence. Cigarettes could have saved a life.

We could do this all day. It starts to take the structure of some kind of sick, emotionally detached game, though, something best relegated to an undergraduate philosophy classroom. In reality, a real man died too young. Tautological biology, neoconservative spin, progressive structuralism aside, the causality here is neat and tidy. We know what killed Morlock. It was Corwin, the tragic character holding the knife.

And yet. What is holding Corwin?

At first glance, *Pomacentrus moluccensis* is a relatively unremarkable creature. The lemon damselfish is yellow and small, and it is a fish. You can basically glean as much from its name. Whatever you are picturing in your head right now is more or less correct.

But lemon damselfish, like many animals, have personalities, and for this reason alone, they are in fact remarkable. Individual *P. moluccensis* display consistent fish-to-fish differences in behavior over time. Some are bolder than others: After encountering the threat of a predator, for example, they'll tend to emerge from hiding spots quicker than their kin. They poke their little noses out and brashly enter the world. In human understanding of fish behavior, this flavor of boldness differs from a personality trait like aggression—the tendency of a damselfish to intimidate or attack another. Aggression differs still from a fish's overall activity level, variation in which separates the Olympians from the couch potatoes. Couch-potato fishes! They're all so different. Boldness, aggression, activity rate: Together, these and other personality traits define what it means to be an individual *P. moluccensis*, as opposed to some indistinguishable carbon-copied toy, fresh off the evolutionary assembly line.

None of which is to suggest that personalities are fixed. In 2009, marine biologists in Australia and California, collecting wild *P. moluccensis* from coral reefs, stumbled into a serendipitous observation. The scientists had been interested in quantifying aspects of fishes' individual personalities, and they'd set to work taking careful measurements of the lemon damselfish in their laboratory tanks. The measurements in question were randomized: They probed different fish at different moments throughout the eleven-day observation period, carefully varying the time at which they tested a damselfish's response to a given intervention—like introducing a smaller fish to the tank, encased in a tiny glass jar—and always varying the order of the measurements. They also avoided observing the damselfish near dawn and dusk, when all fish are known to be more active and aggressive. In other words, they sampled their fish mindfully and attentively, as good marine biologists are wont to do.

But when the scientists began to collate their results, they noticed something intriguing. Because of their careful randomization procedure, they had captured personality measurements of the fish at slightly different temperatures. Over the course of the eleven-day study, water temperatures in the tanks—which were housed in a shaded open-air lab—had fluctuated naturally with shifting ambient air temperatures, which meant those random behavioral samples had occurred in a band from roughly 77 degrees Fahrenheit (25°C) to 81 degrees Fahrenheit (27°C). And when the researchers lined up those personality measurements as a function of temperature, a dramatic effect emerged: Every individual lemon damselfish became more aggressive at higher temperatures. When the researchers plopped those runty little outsider fish into the tanks, the damselfish were on average four times as likely to attack the protective glass jars.

Scientists already understood that there was a relationship between temperature and aggression in fish, but the studies that had lent credence to this idea had always taken the form of population-level understandings. The 2009 study, however, led by Judy Stamps at the University of California, Davis, was the first to document this effect in individual fish. Sure, different lemon damselfish appeared to differ in their natural inclinations toward aggression—indeed, this kind of fish-to-fish variation defines the notion of personality differences in the first place. But all the fish in Stamps's study were susceptible to the temperature effect, and they were susceptible to it *in exactly the same way*: In each fish, for every degree the water temperature ticked up, the number of aggressive acts they committed per minute rose at exactly the same rate. It was like you could watch the water boil their fishy blood.

As it happens, this innate blood boiling occurs in an uncountable number of species. Young black widow spiders are more inclined toward sibling cannibalism in hotter environments. Workers of the high-elevation ant species *Tetramorium alpestre*, which form colonies in the European Alps, are more likely to seize and drag workers from other colonies through the dirt at higher air temperatures. Primates too: If you were to go to the Nanjing Hongshan Forest Zoo and observe the ninety captive rhesus monkeys for days on end, as Chinese biologists did in 2021, you would notice that the monkeys start more fights with one another at 80 degrees Fahrenheit (27°C) than they do at lower temperatures. If I were to list the rest of the known examples of this phenomenon here, this book would be twice as long. But I don't need to do that, and you don't need to go to the zoo and watch the monkeys. Go to a baseball game and watch the pitchers.

In 2011, a group of US economists compiled a dataset of nearly sixty thousand Major League Baseball games: every pitch, every outcome. That's 4,566,468 pitcher-batter pairings. The researchers were interested in baseball's Hammurabi code—the phenomenon in which pitchers are more likely to intentionally hit batters with their pitches if a batter on their own team has already been struck in this manner. The practice plays out with surprising frequency. It is common enough to constitute an informal retributive-justice principle in baseball: a strike for a strike. Even so, the decision to retaliate so aggressively—these are ninety-mile-per-hour fastballs, after all—is an important one, not least because the batter who gets hit also gets to jog to first base. So when *do* pitchers choose to hit a batter?

In their study, the economists quote St. Louis Cardinals manager Tony La Russa: "There are so many conflicting emotions when your batter gets hit. Because how do you sort it out? How do you know for sure that the pitcher acted intentionally?" La Russa suggests that the retaliatory decision relies in part on *interpretation*—in this case, on an interpretation of whether another pitcher meant to harm someone on your team. In other words, there's probably some psychology at play here. Seeking to further understand the mechanics of this decision— and tease out what might be modulating this interpretive act—the US researchers attempted to statistically model the pitchers' aggression. They wanted to know what aspects of the game acted on a pitcher's psychology and decision-making. Were pitchers more retributive if they'd allowed more home runs? If they'd thrown more wild pitches? More walks? Was hitting a batter more likely in a later inning?

No. None of those things mattered. But temperatures did. Even on their own, higher temperatures predicted higher rates of deliberately harming batters. More importantly, though, the effect of temperature on a given pitcher's decision became more extreme with the number of their teammates who had already been struck by the opposing pitcher. In other words, it wasn't just that high temperatures appeared to influence pitchers' aggression; it was that the heat amplified the decision *to retaliate.* By way of example, say a pitcher's teammate was hit in the first inning. On a 55-degree-Fahrenheit (13°C) day, the probability a pitcher will retaliate against another batter is around 22 percent. On a 95-degree-Fahrenheit (35°C) day, that probability rises to 27 percent. As the authors wrote: "Heat appears to magnify the response to provocation: Heat predicts retribution."

But you are already familiar with the crankiness of heat, because you have been alive and tried to navigate the world on a miserably hot day. I don't need to explain your own experience to you. That thing you feel, though? The annoyance? The quickness to the trigger? That is a universal, animalistic feeling. It is intimately within you. It is written in your genes; it is part of your evolutionary inheritance.

Heck, it is encoded in literature. Consider: Dostoyevsky's *Crime and Punishment* begins with an axe murder. The first five words of the book are "On an exceptionally hot evening." There, in an only semi-fictionalized St. Petersburg, in 1866, the heat works "painfully on the young man's already overwrought nerves." The heat is within Raskolnikov.

This dance, between internal stress and external pressure, is always a delicate one, and it's important not to overreach here. Heat isn't the only thing that matters for Raskolnikov. It isn't even the most significant. And yet Dostoyevsky cites the sweltering early July weather again and again in the lead-up to the main event. The stench of sweat grounds the reader in dusty streets. You are there, stinking with the student. Without question, Raskolnikov has real moral agency—the heat doesn't make him steal and wield the axe—but it is impossible to ignore the fact that his actions do come abetted by greater forces: "For his mind was as it were clouded at moments and he was almost unconscious of his body." Rereading, I am reminded that Tyson Morlock, Portland's underpass-pool concocter, was slain in a heat wave on July 1. We mustn't read much into this fact, but I think it is okay to open the book and have a look around.

Elsewhere in books, this environmental context—this "threat multiplier," as the Department of Defense might call it—is ubiquitous. Albert Camus wrote of scorching heat, in 1946, of "cymbals of the sun" clashing on the skull before the pulling of a trigger. Listen to the retired insurance salesman in Ray Bradbury's "Touched with Fire." At 92 degrees Fahrenheit (33°C):

Everything is itches and hair and sweat and cooked pork. The brain becomes a rat rushing around a red-hot maze. The least thing, a word, a look, a sound, the drop of a hair and—irritable murder. Irritable murder, *there's a pretty and terrifying phrase for you.*

It's not just the Western canon (even though, writ large, this canonization of the idea is a useful proxy for an understanding of the heat–aggression relationship as intuitive and human and felt). To pull an instance at near random, look to the early eighth century, in the poetry of Wang Wei (王維; 699–759 CE). In his "Ballad of Suffering from the Heat" (Kǔ Rè Xíng, 苦熱行), Wang writes of a scorching red sun filling the heavens (赤日滿天地), of dried-up riverbeds and ponds (川澤皆竭涸)—and of the understanding that when the body suffers from this heat, the mind is not yet fully awake (卻顧身為患, 始知心未覺). We have long known of heat's sapping, because it has been sapping for time eternal.

The sapping is more than correlational, and its implications are far-reaching. The human zoo we tend to call the city is rife with historical evidence. Riots are more likely at higher temperatures, as are incidents of intimate-partner violence and aggravated assault. The associations between heat and aggression have been demonstrated in a small army of cities, from Dallas to Phoenix to Charlotte to Minneapolis. The effect is identifiable from the littlest fish to the hulking person to the neighborhood, city, and nation. The baseball study was never just a baseball study: "We believe that the increased degree of retaliation at high temperatures in baseball illustrates the role that temperature can play in exacerbating conflict," wrote the authors.

It is also illustrative of the role temperature can play in exacerbating aggression of a much more mundane, much more insidious variety.

In 2019, a Harvard economics PhD student named Ayushi Narayan filed a Freedom of Information Act (FOIA) request for a dataset of Equal Employment Opportunity (EEO) charges brought by workers at the US Postal Service. Narayan, a scholar of workplace conditions, had been seeking to understand the factors that influence workplace discrimination and harassment, and she had been stymied by a lack of available data in the private sector. Yet the USPS constitutes an enormous public workforce—effectively the largest—with almost five hundred thousand employees scattered across the country's tens of thousands of post offices. As we say in journalism: If they work for us, they're FOIAble. Narayan's FOIA request was granted, and she received a dataset of a quarter million EEO complaints from 2004 to 2019—roughly three complaints for every hundred USPS workers per year. It was an enormous dataset of discrimination complaints: about the largest collection that might reasonably exist, anywhere.

Narayan wanted to understand what about a workplace and its surrounding socioeconomic context might influence incidents of discrimination and harassment. In other words, she wanted to estimate the effects of workplace conditions and personal backgrounds on the likelihood of someone committing some broadly defined act of discrimination at work. After months of lobbing statistical methods at the quarter million rows of EEO data, in 2022, she published a precise and exacting study in *Proceedings of the National Academy of Sciences*. Heat was there too, like a cancer. All else equal, compared to days with highs between 60 degrees Fahrenheit (16°C) and 70 degrees Fahrenheit (21°C), days with highs above 90 degrees Fahrenheit (32°C) were associated with a 5 percent increase in EEO incidents. Union grievances ticked up by almost 4 percent at higher temperatures. And just as Anthony Heyes had found in the case of immigration judges, the effects of heat followed people indoors: The temperature–harassment link cropped up among mail-delivery staff and mailroom clerks alike. "The fact that I and many others see some sort of effect even in air-conditioned buildings suggests that there's something about immediate air-conditioning that just may not be enough," Narayan tells me. "At that point in time—the time of flipping on the air conditioner—it's probably not enough to fully adapt."

On our video call, Narayan is careful not to sensationalize her findings, namely because at this point, she finds them deeply intuitive. Like other heat–aggression researchers, Narayan has seen so many of these examples that they barely even register anymore. "I don't even know that it changes my general view on climate change or its urgency," she says. "I think that my study is one piece of the puzzle that has already started to become quite clear [and] is not a great picture. It's in line with the same story that we're seeing over and over and over again." She sounds tired—less like the budding star of economics she is, and more like most of the career climate scientists with whom I've spoken over the years.

For whatever reason, Craig Anderson does not sound tired.

A professor of psychology at Iowa State University and something of the modern father of temperature–aggression research, Anderson is another intellectual heir to Norman Mackworth, the British war-era psychologist who measured the effects of heat on cognition. In the late 1970s, while in graduate school at Stanford, Anderson illustrated the effects of heat on the likelihood of civil riots in the United States, showing that this risk increased linearly with rising temperatures. As with much of the research on the connection between heat and violence, though, the paper offered an all-too-misinterpretable result—namely, one that lent itself to a kind of ugly social Darwinism and that fed a colonial appetite for, say, an understanding of people living in the tropics as inherently violent. But Anderson was no climate determinist. He wanted to understand if there was a measurable psychological effect here, operating at the level of the individual. "If this effect is real, what's going on?" he asked rhetorically when we first spoke in 2015. "Are people just more irritable? Can you replicate this in a lab?"

He'd grown up fascinated with Westerns and World War II shows. "Even as a child I was interested in aggression"—the forces that tip the scales toward violence—he wrote in a recent essay. And after he'd grown a mustache and entered the academy, Anderson retained the interest. He wanted to see how heat changed people, on their own terms, in their own latitudes. So he did what any good twentieth-century psychologist with access to a university campus might do: He began packing people into research labs, tinkering with the thermostat, and watching what happened.

At first, the research yielded little. When there's a kerosene heater sitting in the lab, volunteers catch on to the ruse, and they wind up more suspicious of the experimental paradigm than they do immersed in it. "If your research participants know what you're doing, they start to behave pretty weirdly," Anderson reflected. He needed cover. Fortunately, it was the 1980s. What better cover than failing infrastructure? He started running his experiments in a deteriorating office complex. "With a set of crappy cubicles in an old building, participants could reasonably believe it was just the building's heating system acting up," he told me, referring to a series of studies his team conducted in that period. "They couldn't believe that we had any control over the temperature." With his research volunteers bright-eyed and unsuspecting, the aggression effects popped up like dandelions.

For the following three decades, Anderson would champion this genre of psychological research, always aiming to winnow closer to a behavioral understanding of *how* heat might influence aggression. What was it about the temperature that could provoke such extreme behavior in not just spiders and ants, but also our own species? We like to imagine ourselves as transcending the weight of nature. Craig Anderson knew it wasn't true, and he wanted to be able to explain why.

By the 1990s and many overheated volunteers later, his work had begun to point to what he called an overinterpretation effect. It wasn't that Anderson's research subjects instantly transformed into violent machines at higher temperatures—although, he notes, their feelings of hostility often did increase—but rather, it was that extreme heat more easily lent itself to escalation. Anderson observed, for example, that his subjects were more inclined to interpret otherwise emotionally benign videos as depicting aggressive behavior at higher temperatures. It is this overinterpretation—and the resulting impulse to respond in kind—that lends itself to violence and aggression. In other words, it's about impulsivity and decision-making. "That's what we think is happening in the real-world data at a psychological level," he told me. "That's how bar fights turn into shootings in the parking lot."

The 2011 study of baseball pitchers, with which Anderson was not involved, makes sense in the context of the overinterpretation theory. Recall that the main effect observed by the economists was *retaliatory* in nature. Most sensitively, it depended on what pitchers believed about their opponents' actions. If they were more certain that other pitchers had acted out of malice—if they *interpreted* the other team's actions as aggressive—they were more likely to lash out in kind (and, indeed, even more likely to do so at higher temperatures). This type of behavior is exactly what the overinterpretation theory would suggest. Psychologically, it suggests that higher temperatures prime people to understand one another's actions as more aggressive. And then they're more likely to reciprocate.

In retrospect, researchers had already begun to understand as much by the late 1980s. In 1986, for example, Arizona State University psychologists Douglas Kenrick and Steven MacFarlane turned their attention to drivers in Phoenix, seeking to apply field methods of the late 1960s to the temperature–aggression problem. "We have all probably witnessed normally mild-mannered individuals react to a seemingly minor traffic frustration by leaning on their horns, displaying an isolated middle finger, and letting loose with a barrage of verbal abuse and profanity aimed at a fellow driver," they wrote. It's the kind of outburst the researchers considered to be "ecologically valid"—that is, it didn't depend on artificially stuffing people into an overheated laboratory. For sheer amusement value, I reprint the methodology from the resulting "Ambient Temperature and Horn Honking" study here:

Sessions were conducted on 15 consecutive Saturdays (between 11 a.m. and 3 p.m.) during the months of April through August. The site chosen for the study was a traffic light controlling the exit from a residential tract in an outlying Phoenix metropolitan area. The intersection had a one-lane exit from the residential area, making it impossible to pass on either the left or right of an obstructing vehicle. The green light was set for a 12-second period.

A female confederate, driving a 1980 Datsun 200SX, positioned her vehicle near the target intersection. When the light turned red, she moved her car to the head of the intersection and waited for a subject to pull in behind her. The confederate then waited for the light to turn green and remained stationary throughout the 12-second course of the light. The confederate was instructed to keep still, with her eyes forward, car in neutral, foot off the brake, and her hands on the steering wheel. Once the green light had changed, the confederate made a legal right turn on the red light.

Oh, to be a psychologist in 1980s Arizona. An observer was positioned in the bushes with a notebook: He recorded the number of honks, the duration of each honk, and how long it took for the first honk to emerge from the driver trapped behind the idling Datsun. Back in their offices, Kenrick and MacFarlane matched the honk data with a feed of contemporaneous temperature and humidity recordings they'd nabbed from the university's geography department.

As in the baseball study and Ayushi Narayan's research on postal workers, Kenrick and MacFarlane observed a convincing relationship between heat and aggression. Not only did drivers honk more frequently at higher temperatures; they also spent more time leaning on their horns. At temperatures over 100 degrees Fahrenheit (38°C), a full third of honkers honked for more than half the time the light was green. Nobody did so below 90 degrees Fahrenheit (32°C). Furthermore, the effect was most pronounced among drivers who had their windows rolled down. Presumably, wrote the authors, those cars lacked air-conditioning, and their drivers were more readily exposed to the sweltering outdoors.

Importantly, the maddening driver of the Datsun never changed her behavior. Unwavering as glass, smirking perhaps, waiting for a honk: She did the exact same thing for everyone who lined up behind her. Craig Anderson, who provided feedback on the draft study, would argue: At higher temperatures, the fleet of angry honkers interpreted her actions as more annoying.

Overinterpretation isn't enough for neuroscientists craving a mechanistic understanding of the heat–aggression effect, though. For researchers interested in brain networks and circuits, the psychological explanation looks more like rearranging deck chairs than it does like an answer. It's like the problem with Tyson Morlock and the knife that killed him. Or was it his friend? Sure, okay, we observe more aggressive acts with extreme heat, because extreme heat provokes the overinterpretation of otherwise benign acts. But why does heat provoke overinterpretation?

There is a short, unsatisfying answer here: We don't know. We just don't. But in 2017, a group of Finnish researchers came as close as anyone has to fitting the keystone into the arch.

Jari Tiihonen, a clinical neuroscientist with affiliations at the University of Eastern Finland and the Karolinska Institutet in Stockholm, was perturbed by the information gap, not least because the question of whether climate change will cause an uptick in crime and conflict is rightfully controversial. Surely someone had at least *pontificated* as to why extreme heat leads to more aggressive behavior, neurobiologically speaking. As far as he could tell, though, nobody had. The abstract of Tiihonen's team's subsequent study is written with a hint of near frustration, arguing that "no causal neurobiological mechanisms have been proposed for the putative association between ambient temperature and aggressive behavior." That's about as feisty as contemporary clinical neuroscience publications get. He and his colleagues sought to rectify this problem.

Their approach was relatively straightforward. The researchers began by obtaining a dataset of violent crime incidents from the Statistics Centre of Finland—a half million instances of homicides and assaults occurring between 1996 and 2013. They correlated monthly violent-crime rates with monthly average temperatures (stratifying the data by month to avoid any effects of seasonality), and noted a robust relationship between heat and violence. In particular, ambient temperatures explained a full tenth of the variance in the country's rate of violent crime over the time period the dataset covered. "This was not very surprising," Tiihonen told me, especially "since this association has been common knowledge among police officers." There's an old saying among Finnish cops, he said: "Cold weather is the best police." This much was already known.

But then the scientists went a step further. They started chasing serotonin.

Early in their neurobiological investigation, Tiihonen and his colleagues had noted that suicide rates varied not only seasonally—as studied by other researchers—but also with meteorological variables like temperature, humidity, and daily levels of sunlight. Neuroscientists studying suicide (in the absence of any climatic line of questioning) had identified the brain's serotonin system as wildly important in regulating violent behavior. Furthermore, tracing out the third leg of the stool, Tiihonen and his team observed that meteorological variables had also been previously understood to act on serotonin function in the brain.

Their evidence for this last observation stemmed from neuroscientists studying depression. Many contemporary antidepressants, for example, are known as selective serotonin reuptake inhibitors (SSRIs), because they prevent serotonin molecules from being reabsorbed into the neurons that release them. If something like major depression is, in part, a function of low serotonin transmission, giving the molecules more time to be caught by the neurons that are supposed to catch them ought to help ameliorate the condition. But it turns out, wrote Tiihonen and his team, meteorological variables *also* affect the ability of SSRIs to latch on to the proteins that reabsorb serotonin—and, by implication, *must* act on the brain's serotonin system. There were the building blocks of a hypothesis here: "These findings imply that ambient environmental conditions modulate serotonin (5-HT) function, and that changes in 5-HT function lead to variations in human behavior," they wrote. "However, to our knowledge, no associations have been reported between seasonal variations, violent offenses and serotonergic biomarkers." It was time.

James Callaway, the senior author on Tiihonen's study, had conveniently already compiled a dataset of relevant biomarkers. Back in 1997, he'd led an investigation into serotonin and violence, and as part of his research had collected blood samples from violent-offending patients at a high-security state mental hospital in Kuopio, Finland—as well as from healthy controls who lived in or near the same town. He'd taken the samples at monthly intervals and measured the efficacy of SSRI binding to relevant blood proteins. Monthly measurements, serotonin biomarkers, violent versus nonviolent subjects. All that was left for Tiihonen to do was dig up some weather data from the same time period and stack Callaway's dataset on top of it.

A compelling phenomenon emerged. Warmer weather appeared to downregulate serotonin function. Densities of the serotonin transporter protein that Callaway had measured in patients' blood—the protein responsible for reabsorbing serotonin into the neurons that just released it—were inversely correlated with both high temperatures and higher rates of violent crime. The correlations were highest among violent offenders, with densities of the protein sampled from their blood explaining 39 percent of the variance in the country's violent-crime rate. Importantly, the results were almost as stark for the nonviolent control subjects, leading the authors to ultimately conclude that "the seasonal variation in violent offending is influenced by natural fluctuations in the serotonergic system, and that a 2°C increase in ambient temperature could increase the violent crime rate by more than 3 percent if other contributing factors remained constant."

A little bit of heat, a little less serotonin transmission, a little more impulsivity, a little more violence. A testable theory; a measurable effect.

But a 3 percent spike in violent crime doesn't sound so bad. Does it? In terms of their magnitude, Tiihonen's results are consistent with economists' projections of the effects of global warming on violence. He and his colleagues cite the economist Matthew Ranson, for example, who in 2014 published a paper estimating that climate change will increase the rates of murder and aggravated assault in the United States by a little more than 2 percent. The Finns offered the comparison as means of suggesting the validity of their results. What they didn't do was repeat the absolute numbers that Ranson published. Here they are: "Between 2010 and 2099," he wrote, "climate change will cause an additional 22,000 murders, 180,000 cases of rape, 1.2 million aggravated assaults, 2.3 million simple assaults, 260,000 robberies, 1.3 million burglaries, 2.2 million cases of larceny, and 580,000 cases of vehicle theft in the United States."

Set the neuroscience aside for a moment. Grappling with the climate problem means reflecting deeply on these numbers. Don't pay attention to the actual values; Ranson's estimates are some among many, and nobody can divine the future with that degree of accuracy.

But pay attention to the number of zeros in each figure. Stack up tens of thousands of Tyson Morlocks in a mass grave. The Intergovernmental Panel on Climate Change doesn't usually illustrate as much in the charts and graphs accompanying their reports on the issue. Maybe they should.

That a couple more ticks on the thermostat can make us breathe fire—it's a bizarre bit of psychology, like how time can collapse when viewed down the wrong end of the telescope. Surely your friend only moved to Austin yesterday. Oh, it has been seven years. Time compression is a trick: one of those subconscious phenomena that must have some sensible evolutionary rationale but of which the experience feels uncanny nonetheless. The temperature–aggression link is like that. It is scary to notice the brain at work in this way. It calls into question the degree of behavioral agency to which we have access. It removes some blocks from the Jenga tower of free will.

When does the tower collapse? Studies have illustrated that neurological signals of intent—to decide, to act—can be detected in the brain long before study participants report ever being aware of those intentions. They felt they had free will, as we all do. They were choosing to move their hands. But if you were looking at a computer readout of their brain activity, you could have told them what they were going to do before they decided to do it. Not *much* longer before; these kinds of studies only report effects on the order of 350 milliseconds or so. But in that third of a second, a subconscious decision signal arises from the depths before that decision is consciously made. Yes, it feels like a decision. Was it? Or is every decision merely a perceptual experience of a prepackaged subconscious plan?

Those kinds of questions will make your brain leak out of your ears. And I would humbly submit that they don't matter too much. We still experience the feeling of free will. Detecting the neural signature of a decision before someone is aware of the choice in question doesn't negate the fact that they still *felt they made the choice.* That this feeling arises at all ought to be enough to justify living with integrity and moral responsibility. In other words, just because free will as colloquially understood may not exist doesn't mean hedonism is automatically back in fashion: You still *feel like you're choosing* when you choose to act with cruelty and excess. It's true even if you push over the Jenga tower and run up the flag of nihilism and proudly declare nothing matters. You're wrong. Compassion matters, because you can choose to act compassionately.

But the temperature–aggression link steals some of this choice. It is a thief, and it robs us of compassion. Before Raskolnikov commits the murder, Dostoyevsky says of him, "He thought of nothing and was incapable of thinking; but he felt suddenly in his whole being that he had no more freedom of thought, no will, and that everything was suddenly and irrevocably decided." It's like this when heat takes over. When the tendrils of temperature reach in and take the wheel, you don't feel like you're choosing. Not as much, anyway. By definition, impulsivity is the opposite of thoughtfulness. When heat takes its place in your brain, it is pushing you out.

So let's stop beating around the bush: Did extreme heat kill Tyson Morlock, the man stabbed by his friend in Portland? Harking back to the free-will digression: If the answer is yes—if we, societally, agree to assign some degree of causality to heat's relationship with violence, we also have to agree that we are removing some agency from the person holding the knife. How *much* agency remains an open question. On an extreme end of the spectrum, in the face of violent acts on a hot day, one can imagine a future courtroom entertaining something analogous to a temporary insanity defense. An attorney wheels out an easel and points to a chart from one of Craig Anderson's papers. The jury observes the data. They observe the defendant. Temporary hotheadedness. That guy wouldn't hurt a fly.

Would we ever buy the ethical validity of this defense? Would Morlock's family? I doubt it.

And I question whether we ought to. There's room for a broad tent of causality here. We don't need to sacrifice the principles of accountability to acknowledge heat as a risk factor for violence. We can understand the heat–aggression link as one operating across whole populations, at the margins, without caving to the base whims of climate determinism. The effect is real, and it is scary, but moral responsibility need not evaporate in a heat wave. Knock-on behavioral effects of rising temperatures—dampened cognition, juiced-up impulsivity—are underexamined reminders of the gravity of the climate problem. They ought to remind us of *its* stakes. Did extreme heat kill Tyson Morlock? No, but with a warming globe, more will die like he did.

Paradoxically, researchers' findings like Anderson's and Tiihonen's push back against this suggested inevitability, though. If we really can consider the temperature–aggression link to effectively constitute a problem of impulse control, and if we can understand some of its roots to be stuck somewhere in the thicket of the brain's serotonin system, all of a sudden we're talking about a neurobiological model of behavior that's ever-so-slightly easier to wrap our heads around. In fact, it's a model that already exists.

Serotonin was first properly characterized in the 1930s by Vittorio Erspamer, an Italian pharmacologist who found it in the enterochromaffin cells of the gut. It wasn't until the 1940s, though, in the burgeoning era of psychopharmacology, that the molecule—tentatively called enteramine—was detected in the brain. These were the days of new frontiers. Was it a neurotransmitter too? A Cleveland biochemist soon isolated the compound from blood serum and endowed it with the name "serotonin": a fusion of "serum" and the Greek "tonos," tension, a nod to its effect on blood-vessel tone.

Through the 1950s and 1960s, early neuroscientists delved into serotonin's role as a signaling chemical in the brain; its relationship with mood and mental health began to unfurl along academia's forest floor. The synthesis of SSRIs in the 1970s marked a turning point, offering new hope for treating depression and anxiety disorders. Prozac became a household name. A century since Erspamer's work on enteramine, serotonin is now as close as neuropharmacology has to a celebrity. Popular culture has somewhat corrupted the nuance of its explanatory power, but neuroscientists today take its complexity seriously.

In particular, researchers of attention deficit hyperactivity disorder (ADHD) have long studied the link between the serotonin system and impulse control. Tuck in to chapter 31 of 2020's *Handbook of Behavioral Neuroscience*, for example, and you'll find eighteen tidy pages detailing decades of research on the manifold relationships between brain serotonin transmission and impulsivity, compulsivity, and decision-making in animals and humans. (Don't do that; I'm about to tell you what the chapter says.) Over the years, write the Cambridge authors, a basic finding has been replicated over and over again: Serotonin depletion heightens an organism's sensitivity to negative feedback. That feedback can take myriad forms, but a core truth of brains like ours—and those that look somewhat like ours—is that decision-making in the face of discomfort appears to be modulated in part by the serotonin molecule. Serotonin depletion reliably suppresses an organism's ability to respond appropriately in situations that demand waiting or patience.

Certainly, the molecule is finicky. We don't have all the answers yet in the case of the temperature–aggression link. When Tiihonen and his colleagues compared violent offenders to control subjects in their study, for example, they noted that the former group had *higher* baseline densities of the serotonin transporter protein—but the main finding of the study had suggested an incongruent effect; namely, that as serotonin activity *decreased* with higher temperatures, violence increased. Other researchers have since noted that *high* serotonin transporter levels do indeed appear to explain impulsive aggressive behavior. In an attempt to square the circle, Tiihonen and his team offered that it's perhaps the month-to-month change in serotonin transporter levels that matters, not the absolute amount of serotonin activity per se. Other scientists have argued that with more serotonin transporter proteins come fewer serotonin molecules floating around in the gaps between brain cells. The serotonin transporters scoop them up, after all—that's the action that SSRIs block in the first place. A tendency toward impulsive violence might be characterized by higher levels of the serotonin transporter. It might be characterized by rapid drops in that same protein. Perhaps the story has more to do with the amount of serotonin hanging out in the synapses between neurons, and the transporter protein is just a proxy for something more fundamental. In any case, the heat story isn't so simple.

There may be further clues to Tiihonen's puzzles to be found in impulsivity research, though. Just as in the case of heat and cognition, some prominent ADHD research on impulse control points to impairment of the cingulate—the brain area named after a belt. In ADHD, the entire cingulate cortex exhibits structural and functional differences, relative to brains of people without ADHD—including, sometimes, the presence of literally smaller anterior cingulates. Recall, from heat and cognition research, that as people are exposed to higher temperatures, the anterior cingulate appears to disconnect its activity from that of other cortical areas, including those responsible for executive function. ADHD researchers have noted the same effect in the case of impulse control: While children with ADHD have strongly interconnected amygdalae, these connections appear to come at the expense of connections between the cingulate and the prefrontal cortex. As Boston University School of Medicine neuroscientist Brent Vogt wrote in a recent review for the *Handbook of Clinical Neurology*, "ADHD children/adolescents respond impulsively to the significance of stimuli without having cortical inhibition."

Where does this research leave us? Should everyone start taking Adderall because the world is getting warmer? Far from it. (Besides, because neuroscientists have a rat model for everything, we know from studies of impulsive rats that different ADHD medications act differentially on impulsive behavior, so Adderall might not even be the best bet.) Understanding the interior mechanisms of aggression via studies of impulse control helps us construct a model of what's going on when these behaviors are provoked by heat. Understanding serotonin's role in this interplay allows us to better think through a calculated response.

Drugs are indeed one tool amid a broad tool kit of approaches to managing impulsivity. Presence is another. Researchers have shown that mindfulness interventions—even in middle schoolers—improve ADHD symptoms and the ability to reflect during the kinds of moments that tend to produce impulsive responses. Spend five minutes on any online ADHD advice forum, and you're likely to get a sense of what the rest of the tool kit looks like from the perspective of those who think about impulse control the most. This is expertise born of experience. And people living with and treating ADHD say: Practice self-compassion. Identify the triggers and environments that spur your impulsivity, and then try to minimize your exposure to them. Seek to understand how impulsivity functions. Check in with yourself, and identify the emotions underlying your impulses. And, again and again and again, the advice of people with ADHD goes, consider the future. Nurture a forward-looking mindset. Try to feel a future you. How are you going to feel in the face of this future challenge? How are you going to feel about its various potential outcomes? How are you going to feel if you act on your impulse? It's an approach to impulsivity that's rooted in a mode of present-selflessness.

Which means there's another lesson buried here: Self-control is less about impulse control in the don't-take-the-marshmallow sense than it is about framing decisions in terms of how our future selves will feel. And that means treating our future selves less like strangers. We would do well to apply the same advice to the climate problem.

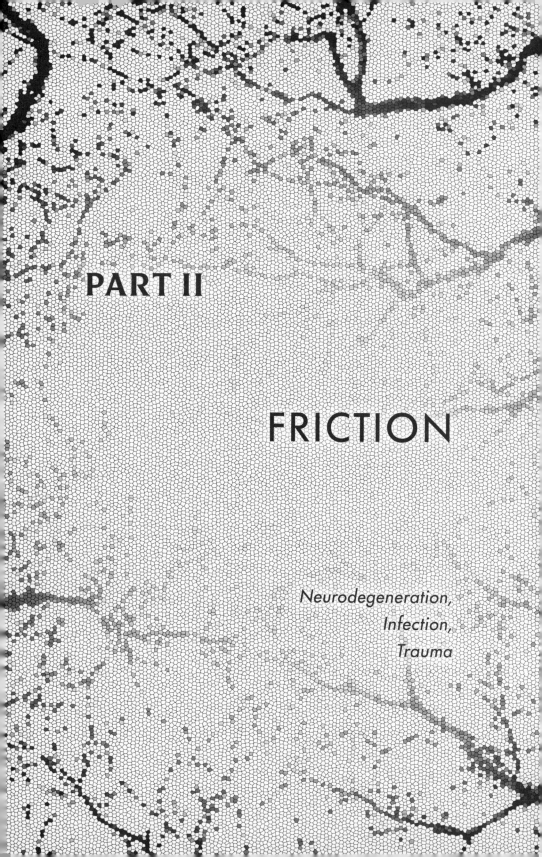

PART II

FRICTION

*Neurodegeneration,
Infection,
Trauma*

4

BLOOM

There was surely nothing to indicate at the time that such evidently small incidents would render whole dreams forever irredeemable.
  —Kazuo Ishiguro, *The Remains of the Day*, 1989

You carry Mother Earth within you. She is not outside of you.
  —Thich Nhat Hanh, interview with Jo Confino, 2012

I t was a little like online shopping, in that there was a courier and there was a package. But David's box was special—more matryoshka doll than Amazon detritus. When it arrived, the outside of the parcel was stamped with diamond crests indicating United Nations codes 1845 and 3373: CARBON DIOXIDE, SOLID; BIOLOGICAL SUBSTANCE, CATEGORY B. He slid open the top in a practiced motion and found the cooler he was expecting inside. A screaky insulated foam lid and a caricature of chemistry—he opened the cooler to a puff of fog. Next, dry ice and biohazard bags, each with a specimen label: IFAW 12-228 Dd, IFAW 12-223 Dd, IFAW 12-200 Dd. He unsealed a bag to find a smaller bag inside. There. Pulling out his prize, David peered through the skin of the final nesting doll. One hemisphere of dolphin brain, frozen in time.

David Davis, a neuropathologist specializing in toxicology at the University of Miami's Brain Endowment Bank, is the kind of man who receives brains in the mail as a matter of professional obligation. The dolphins', seven in total, were from Cape Cod, and in Davis's lab, they joined seven more of a different species from Florida. Each of the fourteen brains had been split along its midline, with one hemisphere from each animal frozen at –112 degrees Fahrenheit (–80°C) and the other soaked in a formaldehyde-based preservative called formalin. The frozen halves were destined for toxicological analysis; the others would be shaved down to slices thinner than red blood cells, mounted on slides, and examined for physical abnormalities.

The brains were marooned there in a Florida lab because between 2005 and 2012, the dolphins to which they'd previously belonged had been found marooned on shores in their home states. It made them ideal research candidates. Dolphin stranding is a curious phenomenon: Of the roughly two hundred thousand bottlenose dolphins living on the Atlantic coast and in the northern Gulf of Mexico, for example, only around eight hundred are found stranded each year. But as whip-smart apex predators, dolphins are considered sentinel species—canaries in the proverbial coal mine. Monitoring and studying stranding events can help scientists wrap their heads around environmental and climate trends that might be relevant not just to marine health but to our own.

"Stranded" is a crafty euphemism, in that it can refer to marine animals found dead or alive—the unifying principle is that they are unable to return to their natural habitats. Sometimes strandings stem from boat injuries or oil spills or entanglements with marine debris. More commonly, though, they're the result of disease or habitat degradation. In 2018, as Davis and his colleagues were submitting the results of their dolphin investigation for publication in a scientific journal, more than 180 bottlenose dolphins began to wash ashore in Southwest Florida—the kind of mass die-off the National Oceanic and Atmospheric Administration (NOAA) would later refer to in a postmodern tone as an "unusual mortality event." A subsequent federal investigation pointed to toxins secreted by *Karenia brevis*, a species of alga that can bloom into colonies the color of dried blood. The compounds produced by these red tides, called brevetoxins, are released when the algal cells burst.

Like Josh Graff Zivin and his interest in the poison of heat, Davis and his fellow researchers were interested in chronic exposures, not unusual mortality events. Rust-colored water that smells like rotting fish is relatively easy to point to as a potential toxin. But not all killers look like blood. Outside of eye-catching hazards like red tides and oil slicks, strandings can present a mystery to science. Some researchers have suggested that dolphins strand themselves when making navigational errors. Shallow, sloping, sediment-rich coastlines, for example, might muck up a dolphin's ability to echolocate, causing them to become confused and run aground—especially if they're used to swimming in deep water. And yet the one popularly held truth about dolphins is that they're supposed to be intelligent. Why would a dolphin ever mistake a beach for its home?

For Davis, the line of questioning had begun a few years earlier with a different box. This one had contained brains belonging to vervets, small gray-haired monkeys that live mostly in southern Africa and the West Indies. The box's mailer, an ethnobotanist named Paul

Alan Cox from the nonprofit Brain Chemistry Labs in Jackson Hole, Wyoming, had asked for Davis's expert opinion on the tissue. Cox's solicitation was, basically: *Anything look weird here?*

Davis had written down everything he could see. A frequent observer of the effects of air pollution on the brain, he was accustomed to the subtle changes that tiny alien particles can spur in the organ. But the stained slices under his microscope didn't look subtle: The tissue looked like a Jackson Pollack painting. It was startling. He knew there was a toxin involved; Cox wouldn't have contacted him were there not. But from a pathological standpoint, the brains didn't look poisoned per se. This wasn't necessarily a story of inflammation. Instead, there was a widespread structural problem: plaques and tangled clumps of protein in the gray matter—the kinds of trail markers that only appear on the path to some kind of neurodegenerative condition.

And so, despite knowing it was impossible, because monkeys don't *get* what he was about to propose, he wrote back to Cox: "Half of these animals look like they have Alzheimer's disease."

From the sky, a cyanobacteria bloom looks like an aurora: electric wisps over dark water. Known colloquially as blue-green algae, cyanobacteria—unlike *Karenia brevis*—are not algae at all, but proper nucleus-free organisms comfortably at home in the bacteria kingdom. It is difficult to overstate how old, evolutionarily speaking, the phylum is. When the first cyanobacteria evolved around 2.7 billion years ago, there was virtually no oxygen in the atmosphere. Earth was mostly shallow ocean and off-gassing volcano; rocks were still a relatively new concept.

Cyanobacteria changed everything. Previous life-forms were microbial foragers, breathing in compounds like carbon dioxide or iron for energy or implementing a rudimentary form of photosynthesis that produced sulfur as a byproduct. But cyanobacteria solved a great mystery of biochemistry: They figured out how to cleave water. The breakthrough afforded a more efficient means of harvesting energy from the sun, and it proved revolutionary. Like sunlight and carbon dioxide, water was everywhere. The princes of the Archean would never go hungry.

For hundreds of millions of years, cyanobacteria fed and multiplied. They formed colonies and evolved filament-like multicellular species. They congregated in enormous floating mats an inch or more thick, collaborating with sulfate-breathing bacteria and stitching their living islands together with carbohydrates and sediment. And all the while, they merrily snapped the bonds of water molecules, converting light into chemical energy to separate the H_2 from the O.

The thing is, though, cyanobacteria didn't have any use for the oxygen. They just needed to nab protons from the hydrogen atoms to power the next steps of photosynthesis. This was good news for oxygen. Liberated from the bonds of water, it was free to do as it pleased.

It was around this moment, 2.4 billion years ago, that cyanobacteria almost killed everything else on the planet. At first, the excess oxygen from the photosynthetic reaction was sequestered by ocean minerals or volcanic gases. But cyanobacteria were prolific feeders, and soon they were emitting more oxygen than the oceans could absorb. For the first time, the gas began to collect in the atmosphere faster than it could be cleared away. It displaced methane. Today, historians of Earth's evolution call the period the Great Oxidation Event. It proved catastrophic for other organisms. Life until that point had, by definition, relied on elements other than oxygen to survive. For these so-called anaerobic life-forms, the new atmosphere was a gas chamber. A few methane breathers survived near hydrothermal vents at the ocean's floor. Much else slowly died.

The atmospheric changes also perturbed the climate. Without the heat-trapping properties of methane in the atmosphere, the Earth cooled to a snowball. There were glaciers in the tropics. For millions of years, the equator was as cold as Antarctica is today. All because something had figured out a new way to eat.

They take and they give. During the deep winter, some life did persist; the building blocks were still there. By the time volcanic carbon dioxide ultimately greenhoused the atmosphere and thawed the ice age, a slate had been cleared for the evolution of organisms that knew what to do with oxygen.

Aerobic respiration, the novel form of metabolism that followed, was nature's greatest invention since cyanobacterial photosynthesis. Newly evolved microbes mined the oxygen-rich atmosphere, leveraging the gas to convert sugars into energy. This innovation paved the road for the evolution of much of what we call life today: the lion's share of bacteria, protists, fungi, animals—us. There wasn't such a thing as a plant until an early microbe engulfed a cyanobacterium and incorporated it into its genome. The internal solar farms that plants use today to photosynthesize, called chloroplasts, descended from this swallowed microorganism. In no small way, cyanobacteria shook the tree of life; endless forms most beautiful tumbled down.

They are still around today, cyanobacteria, splitting water and soaking up sunlight. In autumn and winter, they tend to lie dormant (or, depending on the species, nearly dormant) at the bottom of their favorite water bodies. They can remain this way for months—or, if the water is too chilly, for years. But in the warming summer months, as sunlight penetrates the depths, cyanobacteria begin to grow, with many eventually forming balloon-like gas vesicles to float into the water column. It's here that people tend to recognize them as blue-green algae: Surface-level blooms of cyanobacteria reflect colors corresponding to their photosynthetic pigments, just as plants appear green because of their chlorophyll.

When conditions are right, blooms go bananas. In 2018, a cyanobacteria bloom in Florida's Lake Okeechobee stretched for some 650 square miles: the equivalent of painting half of Rhode Island green. Cyanobacteria communities in China's Lake Taihu can span 900 square miles; they've doubled that record in Lake Erie. In the Baltic Sea, researchers analyzing satellite data have estimated that blooms can blanket up to 75,000 square miles—more than half the size of Germany. As if everything between the Danube and the Elbe just sank to the bottom of the ocean every winter and then came roaring back in June, glinting of emeralds.

Which is all a long way of saying: Maybe it would be strange if they weren't still shaking the tree.

Shortly after David Davis had reported the Alzheimer's-like vervet tissue to Cox, the ethnobotanist had revealed that half of the monkeys had been exposed to an amino acid produced by cyanobacteria. β-N-methylamino-L-alanine, often mercifully abbreviated BMAA, was a suspected neurotoxin, and blinding Davis to the experimental conditions had helped confirm Cox's suspicion—that feeding the monkeys fruit laced with the compound would spur brain damage.

Cox was breathing life into an old hypothesis. In the mid-1960s, an anthropologist named Marjorie Grant Whiting, detailed to Guam by the US Public Health Service, began to investigate the alarmingly high incidence of a neurodegenerative illness in a local Chamorro community. The disease, known as lytico-bodig, was characterized by tremors, paralysis, and dementia—a mosaic of neurological disorders. To Whiting, it looked a lot like lathyrism, a paralytic condition caused by eating a particular type of legume that contains an amino acid called β-N-oxalylamino-L-alanine (BOAA). Aware that Chamorro community members often made tortilla flour from the local cycad plants, she began to suspect that BOAA might be tucked away in the seeds.

She was just barely wrong. Upon chemical analysis, it turned out they contained a compound entirely new to science—something that differed from BOAA by a molecular hair: BMAA.

Over the following decades, researchers would isolate the molecule and illustrate its neurotoxic effects in rats, birds, and eventually primates. It seemed like Whiting's intuition had cracked the case. The only problem was: For the scientists to observe any serious health impacts of the toxin, the animals in question had to be exposed to levels of BMAA far higher than what a person might swallow if they were just making tortillas from cycad flour. You'd have to eat hundreds of pounds of the stuff. By the late eighties, the cycad theory had fallen out of fashion. Funding dried up.

Paul Cox entered the picture around the turn of the millennium, when he and the neurologist Oliver Sacks—who, like Cox, had recently turned his attention to lytico-bodig disease—revived the BMAA hypothesis. In a 2002 paper, the two researchers argued that it was chronic exposure that mattered, not acute doses. Furthermore, they suggested, previous research had failed to consider the whole food chain. Fruit bats and pigs—both of which were staples of the Chamorro diet—ate the cyad nuts as well. Since the toxin could be stored in fats, it was likely that the mammals offered a secondary, magnified vector of exposure.

Cox and his colleague Sandra Banack would confirm their own hunches the following year, detecting enormous levels of BMAA in the local fruit bats. Eating a single flying fox, as the animals were known, would have offered a dose of the toxin equivalent to up to twenty-two hundred pounds of cycad flour.

Alarm bells. The researchers turned their attention back to people. In 2004, working with archived human brains from the Kinsmen Laboratory of Neurological Research at the University of British Columbia, they compared lytico-bodig patients' neural tissue to that of Canadian controls. BMAA was there in the former but not in the latter, whose previous owners had died from conditions like heart failure or lymphoma or pancreatic cancer. There was one exception to the rule: Two of the Canadian brains had come from Alzheimer's patients, and there was BMAA there too. Alarm bells.

By the time Cox and Banack published their Canadian findings, they'd also realized the compound was a cyanotoxin. Cyanobacteria were living in symbiosis with the roots of the cycad trees, pumping out the toxin at minuscule levels and passing it along to the seeds, like a direct mailing of tiny anthrax envelopes. And yet, the tree was nowhere to be found in Canada. What was its toxin doing fifty-six hundred miles from Guam, squirreled away in the brains of Alzheimer's patients? The researchers looked up from the roots of the cycad. Cyanobacteria lived in water on and around every continent on Earth.

Lakes and shorelines. Over the next decade, Cox, Banack, and other researchers launched a series of investigations into cyanobacteria blooms, BMAA, and the compound's neurotoxicity in animals. In coastal labs, the toxin turned up throughout the marine food web. It was there in mollusks and crustaceans and bottom-feeding fish; scientists found it in shark fins and shark muscles. It appeared to be produced by a wide variety of cyanobacteria. And it kept showing up in the brains of people.

David Davis, the neuropathologist from the Brain Endowment Bank, joined the research spree shortly after moving to Miami in 2013. The vervet study would prove jarring, but it couldn't explain how BMAA had wound up in the brains of Floridians, as other researchers at the brain bank had shown a few years previously. Even then, those human brains were archived organs, and they could only tell scientists so much about what was happening in the water then and now. Davis turned back to the food web, in search of a sentinel species—something that would mean something to people. Both dolphins and manatees, for example, have sophisticated nervous systems, live near cyanobacteria blooms, and feed on life that does the same. Understanding their experience might be more akin to understanding ours. ("It's hard to appreciate what BMAA does to, like, salmon," Davis told me.) And ultimately, it was easier to get his hands on dolphin brains.

The resulting study, which Davis and his colleagues published in 2019, reported the detection of BMAA in thirteen of the fourteen dolphins the researchers examined. The one animal that tested negative for the cyanotoxin appeared to have been struck by a boat. Furthermore, the Florida dolphins—which lived in fresh and brackish water, comparatively richer in cyanobacteria—had accumulated much higher levels of the compound in their brains than the Cape Cod dolphins, which tended to spend more of their lives in the deep ocean. And just like the vervets Cox had sent him a few years previously, it looked to Davis like the dolphins were riddled with Alzheimer's-like pathology too.

He doesn't pretend it's a perfect model for people. Dolphins lead very different, much wetter lives. As Davis likes to say, they don't eat steak. "But in the natural dolphin environment," he says, "when these toxins increase, you see corresponding increases in toxicity."

This finding presents a concerning development, because cyanobacteria blooms are blossoming ever wider this century. A recent analysis of lake sediment from Europe and North America, for example, illustrated that 60 percent of lakes have seen an increasing frequency of blooms since the dawn of the industrial revolution, with cyanobacteria outpacing the growth of other microorganisms. The trend has accelerated since 1945. Satellite imagery suggests that 70 percent of lakes around the world have witnessed increased bloom intensity since the 1980s. For neurologists who have begun paying attention to cyanotoxins, it all has the feeling of watching a wildfire jump a highway.

In order to thrive, cyanobacteria need three conditions to be met. First, they need access to the sun's rays. Second, they need an appropriate mix of nutrients. And third, they need warm water. Since the sun isn't scheduled to burn out for another 5 billion years, the first condition is usually satisfied. Thanks to human intervention, the latter two are becoming easier and easier to meet.

Cyanobacteria eat what we don't. Or, more precisely, as agricultural runoff works its way from soils of Midwest corn and Florida sugarcane into gulfs and lakes and estuaries, it nourishes, messianically, whatever it touches. In water, excess nitrogen and phosphorus—essential ingredients in most synthetic fertilizers—offer a nutrient buffet for cyanobacterial communities. Lakes groan under the weight of this runoff: Globally, nearly two-thirds of reactive nitrogen applied to cropland goes unused by the crops in question.

The downstream effects of fertilizer waste are diverse and often unpredictable, but a unifying theme is that microbes win. Each lake and bay holds a distinct chemical balance, and each species of cyanobacteria prefers a different ratio of fertilizer ingredients. Some species can pull nitrogen from the atmosphere, for example, and couldn't give two hoots about how much of the element is in the water. But others proliferate wildly with the addition of extra nitrogen. More often than not, these species produce neurotoxins. Crudely, then: more fertilizer, more BMAA. And if you buy the research of Paul Cox and Sandra Banack and the like: more BMAA, more brain rot.

And, as always, the Great Carbonation Event. Carbon dioxide emissions are dessert at the cyanobacterial buffet. Blooms need warm water to flourish, which a warming climate is happy to provide. Cyanobacteria also need carbon dioxide for photosynthesis. Released from fossil fuels, the gas dissolves in water, where it is blithely apprehended by vibrant communities of blue-green algae.

The temperature and acidification effects are likely synergistic. In 2018, a team of Australian researchers concocted a series of miniature marine habitats, exposing each experimental ecosystem for six months to various conditions climatologists expect oceans to reach by the end of the century. In warm, carbon dioxide–rich waters, cyanobacteria grew to compose 75 percent of the seafloor biomass, compared to 25 percent in control conditions. Bottom-feeders starved, and higher predators suffered in kind. Some 80 percent of energy was lost as detritus. It looked like food web collapse. Except to cyanobacteria.

There's another particularly vicious cycle at work here, in that as a changing climate provokes storms of greater intensity—and greater frequency—agricultural runoff is more likely to be loosed into rivers. On cropland, the use of synthetic fertilizer also spurs the emission of nitrous oxide, a gas with a stronger greenhouse effect than those of carbon dioxide and methane. Nitrous oxide emissions are still well below those of the other major climate-warming gases, but they've increased by more than 50 percent since the advent of the steam engine. Farmers in the United States are deploying nitrogen fertilizers at rates forty times that of their predecessors seventy-five years ago. More emissions, more storms; more storms, more runoff.

It is an old story—3 billion years, at least—a road cut in switchbacks and echoes: the story of the discarded changing the world. I imagine if you were to hold up a microphone to that first cyanobacteria colony and ask it why it was warping the atmosphere, why it was choking out its neighbors to such utter completion, it would say: *We're just trying to grow food and heat our homes.* I might understand.

But I do not have this microphone. I have David Davis on the line, and he does not take the long view. He is looking at his neighbors and looking at his slices and wondering why so many Floridians are winding up with this strange molecule in their brains. In popular science, the 3-billion-year story is usually truncated, boiled down to maxims. It begins with a bloom. It ends with: Don't swim with blue-green algae. Keep your dogs on leash. Eat less shellfish. And even then, says Davis, "the moment you mention seafood in Florida, people just don't want to hear it." The human mind wasn't built to break its habits, much less take the billion-year view. Fine. But Davis is scared, and not because we are not eating fewer mollusks. He is scared because the toxin might be airborne.

Neurodegenerative diseases like amyotrophic lateral sclerosis (ALS), Parkinson's, and Alzheimer's are poorly understood. In part, that's because most cases aren't genetic. In the case of Parkinson's, for example, 85–90 percent of all cases have no obvious genetic cause. The analogous figure for ALS is 90–95 percent. These cases—the vast majority—are known as sporadic cases. They're what most patients have. Aside from the lack of genetic clues, the diseases are also hard to study because animals don't get them. During one of our conversations, David Davis offered that perhaps dolphins that are exposed to BMAA and that develop Alzheimer's-like brain tissue experience memory impairment. But that was speculative. Mostly, neuroscientists who study brain degeneration work on cell and animal models of the diseases that replicate one or more elements of the pathophysiology or symptomatology. But a model is never the real deal. Between 2002 and 2012, clinical trials of drugs for the treatment of the disease had a failure rate of 99.6 percent.

If the lack of clinical progress weren't tragic enough, neurodegeneration is frequently cruel. The most common thing I've heard neurologists say about ALS is that they wouldn't wish it upon their worst enemies. The disease exhibits a callousness that almost defies reality. ALS strikes people in the prime of life and targets their motor neurons. You experience some stiffness and weakness. The weakness is followed by a creeping paralysis. Slowly, you lose the ability to dress, and then to speak, and then to swallow and breathe. The gentlest movements define life. Fewer than 1 percent of people living with ALS experience improvements in their symptoms that last more than twelve months; the main story is one of progressive decline. Elijah Stommel, a neurologist at the Dartmouth Hitchcock Medical Center in Lebanon, New Hampshire, put it succinctly in a recent talk: "Slowly but surely, it just takes everything away from you."

Stommel sees fifty or sixty new cases a year. "It's devastating to watch this repetitive cycle of families and patients going through this terror and tragedy," he said. Despite the cycle, though, Stommel is a rare optimist on treatment prospects for ALS. He thinks good options will become available within the next ten to fifteen years. And that's because neurologists are coming closer to a holistic understanding of neurodegeneration itself—and the manners in which environmental exposures *interact* with other risk factors to produce a disease.

For Stommel, a personal step toward this understanding came from studying the geographic distribution of ALS cases. It was 2008, and a few undergraduates were looking for some letters of recommendation to medical school. Sure, fine, let's see if we can get you some research experience. He was familiar with the burgeoning evidence on BMAA and neurodegeneration, so as an ALS researcher, he thought it would be worth checking to see if there was any relationship between proximity to cyanobacteria blooms and incidence of the disease. He gave the students access to electronic medical records from Dartmouth Hitchcock, as well as other New England community databases of ALS patients, and asked them to plot the anonymized data on a map.

Immediately, a disturbing cluster emerged. Nine cases in the patient database dotted the shore of Mascoma Lake, a western New Hampshire lake known for summer blooms of blue-green algae. Globally, only about two to three people out of every hundred thousand will be diagnosed with ALS each year; the nine cases surrounding the lake corresponded to a rate ten to twenty-five times higher. None of the patients were related to one another, so genetic inheritance wasn't to blame. What was more, the cluster itself was largely concentrated on shores downstream of prevailing winds. Two of the patients, a homeowner and a gardener, it would turn out, had even frequented the same address. "It doesn't mean that cyanobacteria are causing these cases of ALS," said Stommel, "but it does mean there's a risk factor of being near water of poor quality and getting ALS."

In their initial study, Stommel and his students didn't actually detect any BMAA in the Mascoma Lake water. But BMAA is a small, difficult molecule to measure, he said, and it also isn't the only toxin produced by cyanobacteria. The phylum spews out everything from liver-damaging microcystins to hemorrhage-causing cylindrosper-mopsins to the plainly terrifying anatoxin-a(s), a substance produced by the cyanobacteria *Anabaena flos-aquae*, which Stommel told me is something on the order of two thousand times more potent than many agricultural insecticides. Bluntly: "It'll kill a dog in two to three minutes." In all likelihood, exposure to cyanobacteria blooms and the corresponding water and aerosols implies exposure to a cock-tail of cyanotoxins. Throw in other potential environmental toxins—heavy metals, fungal toxins, and the like—and the soup becomes even more intimidating. Plus: We don't know how the ingredients interact with one another once they get inside you.

Today, Stommel thinks the patients in the hot spot identified by the original New Hampshire study were probably exposed to marine toxins as aerosols. Wave action, turbulence, and wind-driven white-caps can all cause bacteria, algal cells, and their associated toxins to become airborne. "Interestingly," he wrote with the other authors of the paper, "we noted several ALS patients who live in close proximity to dams where aerosolization is common."

The aerosolization theory of marine neurotoxin exposure is well accepted, insofar as scientists have confirmed, for example, that bre-vetoxins and microcystins have each been shown to travel via aerosol-ized water. Researchers have shown that aerosolized algal blooms cause locomotor deficits and long-term neuromuscular problems in fruit flies. Others have shown that BMAA targets the olfactory bulb region of mice. In 2014, Stommel turned his attention back to Mascoma Lake. At his side was Paul Cox, the researcher who'd sent the vervet brain tissue to David Davis. Using more targeted methods than Stommel had previously applied, they detected BMAA in carp brain, liver, and muscle tissue. They also found it in air filters.

When I visited Paul Cox's Brain Chemistry Labs in Jackson Hole, Wyoming, the group had just received a shipment of samples from Southwest Florida, where a group of community scientists and advocates under the banner Calusa Waterkeeper had recently begun testing for airborne toxins from cyanobacteria blooms. John Cassani—who himself held the title of waterkeeper at the organization—is an ecologist who'd previously worked as a local government deputy director for water resource management. When he started hearing whispers about aerosolized cyanotoxins, he thought it was worth testing his home of Fort Myers for them. Blooms were frequent in the region; in 2018, a red tide event in Southwest Florida had overlapped with a cyanobacteria megabloom for ten miles. Cassani called the moment "catastrophic." He linked up with Cox, and soon Brain Chemistry Labs had its newest field site. Paul Cox and his colleagues were looking everywhere for the stuff.

Everywhere is a good impulse, because the organisms in question even live in the desert. The thing about cyanobacteria is that they don't need much water to survive. These are spectacularly resourceful species. Crusts and mats of cyanobacteria, for example, blanket much of the deserts of Qatar, where they help bind sand particles together. They lay dormant for much of the year. But when the rains come in the spring, cyanobacterial crusts begin to photosynthesize and grow. This phenomenon intrigued the researchers, because they'd also noticed alarmingly high rates of ALS among Gulf War veterans. What if military traffic in the desert was kicking up cyanobacterial dust? Cox and his team flew to the Gulf, where they would ultimately confirm that the crusts and mats contained BMAA. The toxin wasn't just in lakes. It was in some of the driest regions on the planet. Algal dust. Airborne indeed.

"It would be good if you could stop all military interventions," Stommel offers, only partially in jest, "because the military seems to be a risk factor for ALS. Burn pits, Agent Orange, pesticides: It's easy to make a list of things that are risky and should be mitigated in one way or another." False morel mushrooms (*Gyromitra gigas*) have been linked to clusters of ALS in the French Alps. "Algal Bloom Expansion Increases Cyanotoxin Risk in Food," broadcasted the title of a 2018 paper from the *Yale Journal of Biology and Medicine*. Water-skiing is a risk factor for ALS, presumably because of exposure to aerosols. The story is bigger than the cycad nut.

According to the EPA's unsubtle Atlas of America's Polluted Waters, 218 million people in the country live within ten miles of a polluted water body. In the United States alone, neurodegenerative diseases affect at least 5 million people annually. That number is expected to grow to at least 14 million by 2050. "It should scare everyone," said James Metcalf, an expert on cyanotoxins, as he showed me around Brain Chemistry Labs in jean shorts and Crocs. "It does scare me." The projections mentioned here don't take increased bloom frequency into account, because the professional epidemiological associations that compile the forecasts haven't fully endorsed the cyanobacteria hypothesis of neurodegeneration yet.

Cassani at Calusa Waterkeeper laments the unwillingness of governments to embrace the growing evidence. "Based on what we know, for guidances to only be advisory is astounding to me," he said. For Metcalf, the prospect of climate change bolstering a dementia epidemic is almost comically apocalyptic. He told me, "I just want to be able to look my daughter in the eye and say I tried to do something."

In the summer of 2017, to little fanfare, an obscure United Nations treaty called the Minamata Convention entered into effect. Named for the Japanese city that in the 1950s had played host to the world's deadliest outbreak of mercury poisoning, the treaty set legal limits on the release of mercury into the environment: a Paris Agreement for quicksilver. And by many accounts, it worked. Over the previous decade, the diplomatic and industrial cooperation surrounding the treaty had led to a plateauing of global mercury emissions—and in some regions, a decrease: an all-too-rare success in international environmental policy. The alleged triumph puzzled Amina Schartup, though, because the bluefin tuna she studied kept turning up with more and more mercury in their systems.

A self-described hydrargyrumphile—mercury lover—Schartup is a researcher at the Scripps Institution of Oceanography. Dredging through decades of marine ecosystem data from the Gulf of Maine, in 2019 she identified the culprit that had erased the gains of diplomacy: our warming world. The Gulf of Maine was heating faster than much of the planet, and as waters warmed, predators like bluefin tuna had begun to swim faster and expend more energy. Now the ravenous tuna were seeking out more calories, catching more prey, and, by extension, ingesting more methylmercury: a mercury byproduct that bioaccumulates in fish.

The spike in the substance spells danger for the next link up the food chain (that would be: us). In people, methylmercury is a potent neurotoxin that causes vision loss and speech complications. And more than 80 percent of all human methylmercury exposure comes from seafood.

In other words, the brain-rotting effects of a changing environment aren't limited to cyanotoxins. If that were the case, we'd be getting off too easy. I won't enumerate the entire list of neurotoxins to which you might be exposed on any given day, because I want you to be able to finish this book and still feel comfortable leaving your home. But consider one or two more examples, and trust that there's a point coming.

As water temperatures rise and the seas absorb more and more atmospheric carbon dioxide, paralysis-causing ocean toxins will flourish. Recall that it's not just cyanobacteria that cause harmful algal blooms—consider the case of the red tides provoked by *Karenia brevis*. Some blooming dinoflagellates, for example, release a paralytic compound known as saxitoxin. For them, it's a sex pheromone. For us, it's deadly. With more saxitoxin in the water, more of the molecule will accumulate further up the food chain. Shellfish, including Alaskan clams, are notorious warehouses for the substance, and eating a single clam infused with saxitoxin can trigger what's known as paralytic shellfish poisoning (PSP). I would describe the syndrome and its stakes to you, but nobody does it like my Grist colleague Zoya Teirstein:

> *The truth is that without lab equipment, it's impossible to know when shellfish "go hot." Saxitoxin, the poison that causes PSP, has no detectable effect on shellfish flesh. You can't neutralize it by frying, boiling, or freezing like you would with bacteria. Alaskan subsistence fishers call harvesting your own shellfish "Alaskan roulette"—a nod to the fact that shellfish toxicity can vary beach to beach, harvest to harvest, clam*

to clam. . . . Once saxitoxin enters the human bloodstream, it prevents neurons from firing normally. Signs of PSP tend to develop within a half hour: First, a tingling or burning sensation in the fingertips and lips. Then nausea, vomiting, diarrhea, and finally full paralysis of the respiratory system and death. Just one milligram can be fatal, and there is no antidote.

Saxitoxin is one neurotoxin among many. We know it can prove deadly for people, but in terms of environmental interactions, that's about all we know. The climate-change line of inquiry is new. As the marine chemical ecologist Christina Roggatz wrote with her colleagues in 2019, the fact that ocean acidity and temperature can influence ocean neurotoxin levels "reveals major implications not only for ecotoxicology, but also for chemical signals that mediate species interactions." In other words, while we are only just putting together what's going on down there, we have forecasts of toxic expansion, evidence of acidic, warming seas, and the grim knowledge that many diets depend on the organisms that ultimately absorb the toxins in question.

We also have glimpses into the rest of Pandora's box of horrors. Amina Schartup's research on rising mercury levels sounds quaint next to some of the other contents. Remember how we banned certain flame retardants in the seventies because it turned out they were endocrine disruptors? They're getting pushed back into groundwater sources via sea-level rise, heavy storms, and high winds. Arsenic too. Then there's good ol' domoic acid, another bloom toxin, which researchers showed in 2020 altered how zebrafish swim. Why? It was, effectively, stripping the insulating sheaths from their spinal cords like one might peel a piece of string cheese.

Here's another fun one: In 2021, geologists reported that as the

Greenland Ice Sheet melts, it is apparently releasing mercury into the water at a rate that probably constitutes 10 percent of the global mercury flux through rivers. We didn't know that until the researchers published their paper. It also turns out that mercury, like some kind of cartoonish sludge zombie, is leaking from thawing permafrost.

In short, climate change is mapping an entirely new topography of neurotoxin exposure. And as Elijah Stommel told me, the effects of the compounds are likely synergistic.

Most recently, David Davis has dived into these potential interactions. Turning back to his dolphin brains, he sought to understand the comparative contributions of BMAA and methylmercury to Alzheimer's-like pathology. Upon taking a closer look, he first noted that dolphin brains with high BMAA levels—similar to or greater than those found in human Alzheimer's patients—had up to a fourteen-fold increase in the density of tangled proteins in certain brain regions, relative to dolphins without those levels of exposure. He and his colleagues had appeared to confirm a scaling effect: more BMAA, more Alzheimer's-like pathology. They also identified levels of methylmercury consistent with autopsies of human patients with chronic low-dose methylmercury poisoning. Importantly, the researchers illustrated that dolphin brains testing positive for BMAA *also* reflected problems with the gene encoding a DNA-binding protein common in neurodegenerative diseases like Alzheimer's and ALS. Methylmercury had previously been shown to disrupt the regulation of the same protein. The authors wrote that the evidence "provides a potential mechanism of synergy" for BMAA and methylmercury.

The interactions and multiple exposures at play would appear to hint at a series of causal pathways that researchers have no hope of detangling. But Stommel told me otherwise. The point was never that any one thing caused a disease like ALS. Rather, he argues, multiple or chronic environmental exposures combine with genetic mutations or predispositions to tip the scales toward the disease's presentation and progression. In fact, he can even tell me how many steps it takes: six.

He can say as much because epidemiological statistics on ALS betray a peculiar relationship between incidence of the disease and the age of people who get it. Remember how some cases of ALS are genetic in nature? Just like sporadic ALS patients, genetic patients don't develop the disease until later in life—even though they've carried the relevant mutations since birth. Furthermore, some people with the genetic risk factors in question never develop the condition at all. By comparing the onset of the disease with the theoretical rates of exposure to known risk factors that people *might* experience over the course of their lives, researchers can estimate how many distinct exposures it takes for the toxic and genetic barrage to result in the disease itself.

The idea comes from cancer research. In 1927 in Europe, and by 1940 in North America, oncologists had begun compiling registries of cancer rates to better understand how different populations experienced the diseases—and to be able to track the prevalence of cancer over time. Leveraging these registries, in 1954, British statisticians Peter Armitage and Richard Doll noticed something curious about the age distribution of several common cancers: namely, that death rates of a given age group over the course of a year were roughly proportional to that age raised to some exponent. Today, we know this relationship also holds for the *onset* of many cancers: Annual incidence scales exponentially with age. Armitage and Doll were able to show that this kind of statistical relationship must arise from a multistep process; that is, in the case of cancer, incidence across a population will scale exponentially with age if the disease is a matter of a discrete series of rare cell mutations. By calculating the exponent in question—it was around five or six for many cancers—the researchers could ultimately back out the number of mutations that a given person had to experience in order for the population-level data to bear this relationship.

It turns out you can also apply Armitage-Doll theory to ALS. When you do, lo and behold: six steps.

Most people don't hit six. That's why the prevalence of ALS is so low in the general population, all things considered. But almost certainly, BMAA exposure represents one of the possible steps for those unlucky enough to ingest or inhale it. Indeed, in a 2021 study ranking the ability of a set of environmental factors to causally explain geographic clusters of ALS cases, exposure to the BMAA toxin took home the top honors. No other factor under consideration by the researchers could better explain the spatial patterns evident in patient registries. And as blooms increase in frequency, the number of people taking this additional hit—taking one of their six steps—will only grow.

You've already heard about the Minamata Convention. We regulate mercury because we know what it does to brains. In the case of saxitoxin and PSP, most coastal states in the United States have rigorous testing structures in place to ensure that beaches are closed to fishing when toxin levels rise above a certain mark. We don't have anything like that for β-N-methylamino-L-alanine—and yet we face a rising sea of evidence on the relationship between cyanotoxins and neurodegeneration. A lay observer like myself is tempted to ask, then: When will we finally decide the effect is real? How does any invisible relationship come to be known as fact? I feel like Josh Graff Zivin, studying his new toxicology of heat.

Stommel reminds me that it took decades to prove that smoking causes lung cancer. To that claim, Paul Cox offers: "It has already *been* decades." And at some point, you have to make a call. Science is not a black-and-white painting. We live in the grays.

In the case of smoking, the call began with a drip of scientific findings, a murmur in the medical community. Studies, conducted painstakingly over years, were revealing a troubling correlation between cigarettes and the incidence of lung cancer. The suggestion was initially met with skepticism—even outrage. How could the beloved pastime of millions, a cornerstone of social gatherings, be a silent assassin? To turn this scientific conclusion into policy, the message had to move beyond the cloistered halls of academia and into the public sphere.

In other words, as with most things in life, there was a political angle here too. Critics argue the surgeon general's delay had much more to do with industry lobbying and forgone tax revenue than it did with scientific uncertainty. Regulation can be costly. And for this reason alone, it's hard to get the governor of Florida to care about dolphin brains. As Cox likes to say, "If there's no political will, there's never enough science."

On May 21, 2010, in Madrid, the matador Julio Aparicio stepped into the bullfighting ring at Las Ventas. Dressed in black and gold, he faced a twelve-hundred-pound bull named Opiparo. The animal was strong, but Julio's dagger was sharp, and this wasn't his first corrida. Julio had been in the game for more than twenty years.

His father, a matador with the same name, had been famous the world over in the 1950s. In other words, May 21 should have been another day at the office. Sell some seats, kill the bull, secure the next contract.

Maybe it was the routinized treatment that spelled collapse. Julio the younger—known as Julito—wasn't taking many risks that day in Madrid. He went through the motions. He offered a few flourishes of the cape; he swayed and stepped aside as Opiparo charged. Flip the dagger to the other hand, rinse, repeat. And then he tripped.

On one of its charges, the bull had begun to turn around just after passing Julito's cape, and the man had stumbled over one of the animal's legs. He fell. And as he scrabbled backward on the ground, Opiparo, mere feet away, charged again. The matador's misstep produced what is likely the most famous bullfighting photograph of all time: Julito, eyes closed, lanced by Opiparo's horn—through the neck and out the mouth.

When I met Julito's cousin, Manuel Aparicio IV, he asked me if I had seen the photograph. Manny had a 1950s-era poster of Julio the elder in his home. He recounted his uncle's storied career and, gesticulating in a manner that only makes sense in the context of the abovementioned photograph, described Julio Jr.'s infamous 2010 performance. (Remarkably, after a snappy tracheotomy and jaw reconstruction, Julito survived the episode and was back in the ring a few weeks later.) As my editor will not let me include the photograph in this book, if you are the kind of reader who can tolerate gore, I recommend you avail yourself of your nearest search engine. Few pictures are more humbling.

Manny, the Aparicio I had come to Southwest Florida to see, wanted to introduce me to more than his family's bullfighting history. He wanted me to meet something named ADAM.

ADAM is, effectively, a cooler stuffed with plastic tubing. It is also an acronym: the Aerosol Detector for Algae Monitoring. Manny is its creator. A retired engineer with a PhD in computational neuroscience, he volunteers as a ranger for John Cassani at Calusa Waterkeeper. When he's not on his boat, he's fiddling with ADAM and its clones. When we travel together on a late summer morning to one of his sampling sites—a friend's waterfront backyard—he's so excited to show me the contraption at work that he dips his bare hand into a cyanobacteria bloom to pull it out. "I probably shouldn't have done that," he says with an easy smile.

In the field, ADAM looks roughly like a boom mic and an intravenous drip line attached to a lunch pail. It makes a faint whirring sound, and there are bubbles in tubes. It's sampling the air and the water. Manny estimates you can build an ADAM for about six hundred dollars. This is backyard science.

It is also cutting-edge, world-class science. Nobody is conducting the kinds of trials that Manny and Calusa Waterkeeper are. Preliminary results from their first samples—the fleet they'd sent to Jackson Hole—showed they weren't barking up the wrong tree. BMAA appeared to be airborne, just like other hints of evidence had suggested. No need to scour the Qatari desert: It was right there in their backyards. When David Davis reviewed their initial findings, he told me, "If their data is real, the concentrations they're getting are really scary." He points to other studies showing what he calls "crazy amounts" of BMAA in human olfactory nerve tissue. "There's no blood-brain barrier there. It's really frightening."

Manny, whose home also backs up onto water—a mangrove forest—doesn't worry too much about his own health, despite living in such close proximity to something that's potentially so dangerous. In his telling, he's had a good life. Sold a company; got a couple of degrees; gets to volunteer on the water. And no dementia. He does, however, worry about a younger generation. "It doesn't matter if you eat fish or not," he says. "It doesn't matter if you swim or not. It doesn't matter if you live right on the water or not. The air travels for miles." Like David Davis, on behalf of young Floridians, Manny is frightened by his aerosol data. That's where fear enters the picture.

His fear also stems from the fact that the machinery of public health policy is vast and labyrinthine. There is no single lever to pull, no solitary button to press that can swiftly enact regulatory measures—especially when the subject of scrutiny is ubiquitous. (In the mid-twentieth century, the clouds of tobacco smoke that often filled homes, offices, and public places were as common as oxygen itself.) The course of such action is an intricate choreography of scientific research, public outcry, bureaucratic process, and political will. Policy is a colossal beast, slow to awaken and slower still to move. In the case of cigarettes, the tobacco industry fought back fiercely, leveraging its considerable influence and wealth to sow doubt about the science and delay regulatory action. It took the intervention of Luther Terry, the surgeon general in 1964, to respond to the growing scientific evidence. In a historic report, the advisory committee he convened concluded unequivocally that cigarette smoking was a health hazard of sufficient importance to warrant appropriate action. Regulation began to trickle down.

Manny says he's no Luther Terry. So what is he doing here, anyway? Packing tubes into insulated foam coolers and express-mailing them to Wyoming? Circulating op-eds? Is that enough? By his own estimate, he's just a retiree tinkering in his basement. His uncle was the bullfighter; his cousin is the one with the scar.

But watching him work on the dock that morning, I'm not so sure

I buy the humility. The road to regulation is long and winding. And while policymakers aren't acting with urgency, Manny is.

So it's in this light, and against the very real glare of the sun off the water, that I let his boating vest become a kind of cape to me, and for a moment, in the wind moving through the mangroves, I choose to hear the people chanting for the matadors at Las Ventas. It is worth noting that β-N-methylamino-L-alanine carries the chemical formula $C_4H_{10}N_2O_2$. A skeletal drawing of BMAA suggests a carbon squiggle of a body; the molecule is adorned with two oxygenated horns. If you squint a bit, and you listen for the roar of the crowd there too, it is the constellation hunted by Orion: a tiny Taurus.

5

SPILLING

The great slime kings
Were gathered there for vengeance and I knew
That if I dipped my hand the spawn would clutch it.
　—Seamus Heaney, "Death of a Naturalist," 1966

Corporeal life is indeed difficult.
　—David Abram, *Becoming Animal*, 2010

P hilip Gompf had spent the day wakeboarding with his cousins.
Nothing out of the ordinary there: It was the summer of 2009,
in Auburndale, Florida, and he had just turned ten years old.
What else is a ten-year-old boy supposed to do halfway between
Tampa and Orlando besides jump in a lake with his family?

When Sandra Gompf, his mother, tells the story of Philip's death,
then—the story of what happens after the lake—it is the specific var-
iant of rapidity that is so jarring. One day her boy was there. He was

swimming. The next week, Sandy donated his organs. This ragged celerity is not the same kind that comes with other flavors of tragic accident like plane crashes or falling cranes and the like. In the latter cases, there are awful fireworks, and in an instant, there is smoke. In Sandy's story, there are no fireworks. There is just the water, unmoving.

The trouble had begun five days after the lake, when Philip complained of a bad headache. Sandra Gompf is an infectious disease specialist, and her husband, Tim, is a pediatric hospitalist. They are the kind of parents you want in your court when a bad headache arises. Headaches for Philip were rare, but it was the summer. He didn't have a fever. His neck wasn't stiff or anything, so they were able to rule out meningitis. It was probably just dehydration. They sent him to bed.

The next morning, Philip was hard to rouse. Sandy was already at work. When Tim checked on their son, he noticed, this time, that his neck was indeed quite stiff. He could barely bend it forward. Tim rushed him to the emergency room, at the same hospital where he worked, and Philip got a spinal tap. There was a glaring degree of inflammation; the results suggested a severe meningitis. It was probably bacterial.

But there weren't any bacteria to be found in Philip's spinal fluid. The lab couldn't identify anything but the inflammation itself—the type of inflammation that, as Sandy would later tell the science journalists Erin Welsh and Erin Allmann Updyke, "is associated with death." Within a few days, Philip began to develop seizures. Then he went brain-dead. It was shortly after that moment, three days after his symptoms had first begun, and about eight days after swimming in Lake Arietta, that Sandy and Tim had to make the impossible decision: to take Philip off life support. Which they did.

And then the results of an autopsy Philip's parents had requested during organ donation came back. Disbelief continued, somehow, to mount. Their boy's cause of death had been determined to be an amoebic meningitis caused by the species *Naegleria fowleri*. You perhaps have heard of this organism by its colloquial name: the brain-eating amoeba. Which is not a misnomer.

Naegleria live in fresh water. They often eat bacteria, but they'll also settle for brain cells. The amoeba looks like what you imagine when you imagine an amoeba: blobby, with little protuberances called pseudopods. This form is known as the amoeba's trophozoite state—the state in which it can feed and divide. As a trophozoite, *N. fowleri* also sports little intertube-like appendages. These structures, known as "suckers" or "food cups," are what it uses to nibble on other cells.

Amoebae exist not just as solitary organisms but as somewhat of a metaphor for the fluidity of life itself. They are creatures as nebulous as clouds: unicellular—a single cell embodying the entirety of their existence—but housing a universe of complexity. Amoebae like *Naegleria* are often found in moist environments, from the dew-drenched leaf litter of the forest floor to the depths of the oceans and freshwater bodies like Lake Arietta, even occasionally thriving in the invisible rivulets within the soil beneath our feet. To observe an amoeba is to witness a certain embodiment of flexibility. They move and explore their world by extending and retracting their pseudopods. They engulf their food, enveloping and internalizing their prey. But they also incorporate bits of their environment along the way, experiencing the world in a manner that is uniquely their own. The amoeba holds no solid form. It is a creature of change, its entire body a single adaptable cell that flows around obstacles, encapsulates sustenance, and divides itself to reproduce. It thrives in the boundary spaces—the liminal realms where water meets land and where root tip encounters soil—constantly adapting and molding itself to the contours of its world. "Amoeba" comes from the Greek "amoibe": change.

Naegleria are particularly shape-shifting, even for amoebae: Each one can morph from its feeding form to a flagellated, swimming version of itself. *N. fowleri* are the only free-living amoebae that can swim in this manner. They can swim in lakes, and they can swim in your cerebrospinal fluid. They can hunt. They can search for nutritious meals and the best environmental conditions. And, if the water gets too cold, they can stage-shift again. The dormant form of *N. fowleri*, known as the cyst, can be viable for years. We don't really know for how long *N. fowleri* cysts can actually persist. Nobody has done those types of studies. But research on other free-living amoebae would suggest they're hardy. Scientists have desiccated amoebic cysts. They have irradiated them. They have exposed them to bleach. And still, they can survive up to twenty-five years. Each amoeba is, evolutionarily speaking, operating under the assumption that it is the last of its species. Its job is to survive, and it is very good at its job.

Still, Philip Gompf's tragedy is vanishingly rare. Globally, our best estimates suggest a few dozen people contract primary amoebic meningoencephalitis—the name given to Philip's condition—annually. Researchers studying case reports believe the amoeba has infected a couple of hundred people since the 1960s, when epidemiologists and parasitologists began to put a name and face to *Naegleria fowleri*. Yet for something on the order of 97 percent of them, infection has implied death. It is rare, but it kills with ruthless efficiency.

It is becoming less rare. To live, *N. fowleri* requires a bath of 80 degrees Fahrenheit (27°C) or so, and it grows well at temperatures up to 115 degrees Fahrenheit (46°C). As waters warm globally, more

N. fowleri are waking up. In 2023, a Nevada two-year-old died from primary amoebic meningoencephalitis that the state health department believes he contracted at a natural hot spring. The amoebae live in lakes and rivers and hot springs; in discharge from power plants; in puddles and buckets and poorly cleaned swimming pools; on splash pads and Slip 'N Slides. In their cyst form, *N. fowleri* exist in an effective state of suspended animation, waiting for the right metabolic conditions to descend and shake them to life. And then they start to feed. They are perfectly happy living in pipes connected to tap water.

To be clear, you can't catch a *Naegleria* infection from drinking, and you also can't catch it from somebody else. The amoeba has to enter your body through your nose. (Neti pots are a risk factor.) And if you're unlucky enough to wind up with some of the pathogen in your nostrils, just as David Davis believes of β-*N*-methylamino-L-alanine, the amoeba can pick the lock of your blood-brain barrier. It produces a suite of enzymes that break down your nasal mucus, and it hitches a ride on your olfactory nerves to sneak through the cribriform plate, a thin, perforated section of skull at the back of your nose that connects these nerves to your olfactory bulbs. Once it's in the brain, all bets are off. As a trophozoite, *N. fowleri* will consume red blood cells and whole brain cells by engulfing them. It eats; it divides; it conquers. As it spreads, hosts begin to hemorrhage. Brain tissue dies. White blood cells come to the ill-fated rescue, and severe inflammation increases the pressure inside the skull. Coma follows. Death after that.

Five years after Philip's death, in 2014, his parents threw themselves wholeheartedly into advocacy efforts. A memorial fund had been established in Philip's name, and they ultimately decided the dollars would go to *Naegleria* infection prevention and awareness. As the Gompfs like to say, primary amoebic meningoencephalitis is 99 percent lethal but 100 percent preventable. You just need to keep water out of your nose. The campaign they've founded, called Summer Is Amoeba Season, is meant to signpost the midyear analogue to flu season. They've run billboards. They've helped develop a best-practices treatment regimen. In the summer, their nonprofit passes out free noseclips.

Centuries ago, our ancestors lived in close quarters with animals: sharing their habitats, their resources, and, unwittingly, their pathogens. The signs of this risk were there in the form of pestilence and plagues, yet the connections remained elusive. One of the earliest recorded intimations of a disease passing from animals to humans comes from the works of Hippocrates around 400 BCE, in which he noted how certain maladies appeared to be shared across species; yet the concept was nebulous, more observation than understanding. The dawn of the nineteenth century ushered in a deeper recognition of the animal–human disease bridge. In 1854, the British physician John Snow traced the source of a cholera outbreak to a water pump contaminated by human waste, thereby demonstrating the role of fecal-oral transmission, a pathway that would become all too familiar in what would ultimately become known as zoonotic diseases.

The full gravity of the animal–human disease bridge wouldn't be realized until relatively late in the twentieth century, though. In the 1960s, a series of outbreaks—Marburg and then Ebola—shocked the world into abrupt attention. The viruses, both hemorrhagic fevers with terrifyingly high fatality rates, had seemingly sprung from nowhere. Yet careful investigation showed they hadn't: Lab workers, for example, were shown to have been exposed to tissue from infected grivet monkeys. It became apparent that the diseases were zoonotic. Such grim revelations led to an increased emphasis on the study of animal-borne maladies, and gradually, the intricate story of inter-species disease transmission began to unfurl. Each thread revealed dozens more. Some of the twenty-first century's most devastating diseases were revealed to have animal roots. After its discovery in the 1980s, HIV was ultimately traced back to nonhuman primates.

Here is the view from the contemporary ziggurat. Beyond the neurotoxic effects of environmental flux—beyond BMAA and methylmercury and the like—a changing climate is also bringing people into closer contact with vectors of brain disease. Some of these, like *Naegleria*, are organisms that can infect people directly from environmental sources. Others require an intermediate host.

At least 60 percent of infectious diseases are zoonotic in origin. The corresponding figure for new and emerging infectious diseases is 75 percent. Zoonoses are responsible for nearly 3 million human deaths annually. As deforestation and urban sprawl and the ecosystem shuffles of climate change push people and nonhuman animals closer to one another, it becomes more likely—simply as a matter of billiard balls bouncing off of one another—that people catch diseases from other animals.

The 2014 Ebola outbreak started this way: We logged and mined 80 percent of the habitat of Ebola-virus-carrying bats, which proceeded to wander into a village in Guinea and set up camp above patient zero's home.

As the tendrils of urban sprawl creep ever outward and complex ecosystems yield to cement and lawns, intricate relationships between species are disrupted, with unforeseen consequences. Forests fall; grasslands are razed for farmland or urban expansion; and the myriad creatures who call these places home are evicted from their ancestral lands. Creatures great and small are forced to eke out an existence on the margins, pushed ever closer to human habitats. This enforced proximity between humans and wildlife—this mingling of worlds—sets the stage for the potential transfer of pathogens from animals to humans.

Brain diseases are no different. The brain is an organ. It is like our others in that it is defined by specific types of tissues. Some disease vectors, like *Naegleria fowleri*, are particularly well suited to these tissues. Always in the neuroscience of infectious brain disease, though, pathogens like *N. fowleri* need to actually enter the brain. This is easier said than done. The blood-brain barrier is a complex system of cells that forms an impermeable barrier between the bloodstream and brain tissue. It acts as a formidable gatekeeper, regulating the passage of substances, both beneficial and harmful, between the blood vessels and the brain. Composed primarily of endothelial cells lining the brain's capillaries, the blood-brain barrier is also reinforced by astrocytes—star-shaped cells that envelop the blood vessels—and a flurry of other supporting cell types. In one's day-to-day life, the barrier allows essential nutrients, oxygen, and hormones to nourish and support the organ's intricate network of neurons. In the best of cases, it also blocks toxins and pathogens. (The immense selectivity of the blood-brain barrier can pose challenges to therapeutic drug delivery.)

As we've seen, *Naegleria* has cracked the barrier's code by sneaking its way into the brain via the olfactory nerves. There are other pathogens that take the same route. There are others still that travel in your bloodstream and brute-force their way through the astrocytes and endothelial cells. Many of the carriers of these pathogens are set to flourish under climate change.

There is Japanese encephalitis, a mosquito-borne illness caused by a virus from the Flaviviridae family. Infections from Flaviviridae in their own right, common in East and Southeast Asia, as well as some stretches of the western Pacific like Australia and New Zealand, are often mild. Frequently they are symptomless. But if the virus in question crosses the blood-brain barrier, it can cause the encephalitis implied by the name of the more severe disease. In these instances, fever, headache, and muscle pain give way to delirium, convulsions, cognitive impairments, and paralysis. About a quarter of cases are fatal. As our planet warms, the range of the *Culex* mosquitoes carrying the virus expands.

There is neuroborreliosis, an insidious consequence of Lyme disease. Caused by the bacterium *Borrelia burgdorferi*, the condition can lead to a range of debilitating neurological symptoms, including cognitive impairments, headaches, and facial paralysis. Diagnosis often eludes the suffering, since the symptoms can be mistaken for multiple sclerosis or some kind of viral meningitis. In the case of neuroborreliosis, the primary vectors for transmitting the bacteria to people are ticks—and the relevant arachnids are on the move. As with the story of Flaviviridae, *Culex*, and Japanese encephalitis, as the planet heats up, ticks carrying *Borrelia burgdorferi* expand their habitats and their active seasons. Cases of neuroborreliosis rise in kind.

There is yellow fever, another viral disease transmitted by mosquitoes. Yellow fever is most common in the world's tropical areas. Traditionally understood as a flu-like malaise, the infection can spur severe jaundice, bleeding, and organ failure. But the virus can also infiltrate the brain, where it causes neurological havoc, disrupting the delicate communication between neurons and spurring seizures, delirium, and coma. We have a vaccine for yellow fever, but its provision is spotty. In what you are perhaps recognizing as a theme, rising temperatures and altered precipitation patterns create favorable conditions for both the *Aedes aegypti* mosquito, the relevant virus's primary carrier—it is known as the yellow fever mosquito—and the virus itself. Expanded *A. aegypti* habitats and increased viral replication rates lead to higher infection rates in people.

There is Zika, of terrible head-shrinking infamy, also carried by *Aedes aegypti*. Once considered relatively obscure, the illness and its associated virus—like that which causes Japanese encephalitis, also from the Flaviviridae family—burst into prominence in 2015 when it swept across the Americas, leaving a trail of devastation in its wake. Part of Zika's insidiousness lies in its ability to cause severe birth defects, particularly microcephaly, in infants born to infected mothers. Affected children face developmental delays, motor dysfunction, and hindered information processing, which ultimately impacts one's ability to learn. You already know that warming weather can increase the range of *A. aegypti*. Extreme weather events, too, intensified by climate change, can provide new breeding grounds for the mosquitoes, which flourish in pockets of stagnant water.

There is cerebral malaria, which is not caused by a virus, but is still indeed carried by a mosquito. Disproportionately affecting children in sub-Saharan Africa, cerebral malaria is an infection of the *Plasmodium falciparum* protozoan. It is relentless and deadly. When malaria-infected red blood cells obstruct tiny brain vessels, oxygen-starved neurons suffer, leading to seizures, coma, and death. But cerebral malaria isn't just an infection; it's also an inflammatory response gone awry. The parasite triggers immune cells to release chemicals that damage blood vessels and prompt swelling and hemorrhages. Survivors face long-term cognitive and neurological impairments, including learning difficulties and motor deficits. *P. falciparum* tends to be carried by *Anopheles* mosquitoes, which are due to experience range expansion under a changing climate.

There is Powassan virus, another pathogen lurking in the shadows of the tick world. Powassan infiltrates the body, ultimately attacking the nervous system and causing severe brain inflammation. Victims can experience severe headaches, seizures, and long-term neurological damage. With warming temperatures and expanding tick territories, the Powassan range is broadening.

There are more. I will stop. I know it's getting repetitive.

In the case of *Naegleria fowleri,* Christopher Rice, a parasitologist at Purdue University, suggests that we're already seeing the effects of climate change bear on the amoeba's likelihood of infecting a human host. While cases of primary amoebic meningoencephalitis tend to be most common in the southern United States, for example, "we recently saw cases in Minnesota, and last year, there was a case in Iowa," he says. "There are cases in these northern-tier states." For Rice, who speaks with a thick Scottish accent, cases like Minnesota's suggest that the amoeba is already present in all fresh water. The difference could merely be that as temperatures rise, people—especially children—are more likely to jump into lakes.

But the skew of cases toward children (particularly boys) is a uniquely US phenomenon, he says. In other countries, cases are more common among middle- to late-aged men, who may, for example, spend a disproportionate amount of time in rice paddies or practicing nasal ablution, relative to the US context. Understanding future global prospects for the incidence of *Naegleria* infection will also require a deeper epidemiological understanding of the underlying social factors that bring people into contact with contaminated water in the first place. "The United States just has a weird age distribution," says Rice. "It's mostly young male children. And we just think that it's because they are jumping into the freshwater lakes without any thought or hesitation."

These waters are muddy. There does appear to be a core correlation between water temperature and sightings of *Naegleria fowleri*, but Rice is quick to note that we have to be conservative in our interpretations here. For example, if water temperatures rise too much, they can kill the pathogen. "It's a weird cycle we need to think about," he says. It's true that warmer waters appear to cater to the organism, but to Rice, that dynamic could suggest a range *shift* as opposed to a range expansion. We just don't know. In speaking with scientists about free-living amoebae, I frequently heard how little people know about these organisms. We don't know if this one can adapt to higher temperatures. "The free-livings have to survive or else their species will be wiped out. So their mentality is adapt to survive." Hence the evolutionary adaptation to infect hosts.

As Rice argues, we also just don't have good diagnostic tools yet. Most medical professionals don't suspect *Naegleria* when a patient presents with headaches and a stiff neck. It certainly wasn't the Gompfs' first instinct. "And therefore," says Rice, your average *N. fowleri* infection "just gets misdiagnosed—and then doesn't get treated properly." ("Properly," in this case, refers to a cocktail of antifungal and antibiotic drugs that, if delivered quickly enough, can sometimes halt the proliferation of the amoeba.) When I ask him if we're already undercounting cases, he offers the example of a hospital in Karachi, Pakistan: "There's one physician in one hospital in one city who found twenty-four cases in one summer," he says. Rice suspects we're only catching one-tenth to one-fifth of patients.

One of my least favorite activities is asking parasitologists who think about climate change what scares them the most. Here is Christopher Rice's answer, which feels worth quoting in its entirety: "We really don't have any good drugs against these pathogens. One or two of the survivors who have survived are on complete life support. That quality of life is extremely poor. Picture this: You're outside enjoying yourself and having fun and jumping into a lake or pond not thinking about a microscopic amoeba that's gonna eat away at your brain. You know, that's the terrible thing." The thing is terrible, and we're not ready for it to occur more frequently.

I am picturing Rice's hypothetical, but I am picturing more than it. Beyond the individual, the effects of under- and misdiagnosis ripple outward, pebble-like. Each missed case represents a lost opportunity for understanding, a missed stroke in the detailed sketch of the epidemiology of primary amoebic meningoencephalitis. The absence of accurate counts leaves physicians and researchers navigating in the dark, unsure of the true magnitude of the storm they face. Almost by definition, these gaps hinder the development of effective treatments and prevention strategies. A disease underestimated is a disease uncontrolled, and the opportunity for early intervention—that golden key to many medical victories—is lost.

While emerging zoonoses and brain-eating amoebae tend to offer the most high-profile headlines and shocking epidemics, endemic zoonotic diseases exacerbated by climate change—the stuff we already know about—might be even more insidious in terms of their aggregate toll on human health. As researchers at the Centers for Disease Control and Prevention (CDC) wrote in 2017, for example, "The 2014 Ebola epidemic was responsible for 11,316 deaths and

$2.2 billion in economic losses, whereas each year rabies accounts for ≈59,000 human deaths and roughly $8.6 billion in economic losses worldwide."

Rabies. Huh. I had indeed more or less forgotten about rabies.

But as the CDC suggested, the virus is devastating. Encased within a bullet-shaped envelope, the rabies virus possesses a single strand of RNA. It is typically transmitted through the bite of an infected animal. And once the virus gains entry to a human body, it infects nerves. It races toward the brain, and it hijacks behavior. Formerly known as hydrophobia, the disease infamously causes a fear of water. Perhaps more infamously, though, it turns the docile into the deranged. As infected neurons succumb to the virus, hallucinations, anxiety, and convulsions give way to uncontrolled aggression, foaming at the mouth, and a propensity to bite—an attempt by the virus to propagate its lineage. Rabies can be carried by dogs and foxes and skunks and raccoons and coyotes. It can be carried by bats. In addition to having forgotten about rabies, I realize I had more or less forgotten about bats.

The US Geological Survey is quick to inform me: "Vampire bats do not suck blood—they make a small incision with their sharp front teeth and lap up the blood with their tongue." I do not take much solace in this distinction.

Vampire bats are real; this much I already knew. I did not really know they carried rabies, nor the extent to which the disease tends to be fatal upon contraction. I did not know the federal government maintains a surveillance program dedicated to tracking vampire bats' whereabouts. People work in this program. The United States employs vampire trackers. They have pensions.

The first federal Van Helsing I meet is named Ryan Wallace, a veterinary epidemiologist who leads the rabies epidemiology team at the CDC. His team sits within the Poxvirus and Rabies Branch, which itself resides within the Division of High-Consequence Pathogens and Pathology in the Center for Emerging and Zoonotic Infectious Diseases—one of the centers to which the "Centers for Disease Control and Prevention" refers. He also runs the CDC's World Organisation for Animal Health Reference Laboratory for Rabies. Wallace conducts rabies research in dozens of countries and has participated in the vaccination of hundreds of thousands of dogs. He is the rabies guy.

Vampire bats are not the villains they're often made out to be, says Wallace. They are a study in resilience and ingenuity. Desmodontinae are perfectly adapted to nocturnal life, with bodies streamlined like whispers and wings that beat with a silence that is as awe-inspiring as it is eerie. They navigate a world almost entirely unseen by human senses. It is merely their food—their affinity for the blood of other mammals—that casts them in an uncanny light. Razor-sharp incisors, evolved to precision, pierce skin with a surgeon's finesse. The bitten are often oblivious. The bats' saliva, a cocktail of anticoagulants, ensures a steady flow of blood.

Perhaps even more intriguing is their social fabric. Vampire bats, contrary to their common solitary image, are deeply social creatures. They dwell in roosts bound by ties of kinship and cooperation. Mothers care devotedly for their young; siblings share in duties. They exhibit altruism. They share meals. Their roosts are social networks, built on bonds that endure across time.

Accordingly, the climate theory of vampire rabies looks something like this: Vampire bats are living beings with a complex ecological role; their existence is interwoven into a delicate balance of ecosystems. But with the ebb and flow of climate change, the animals face a double jeopardy. Capricious weather patterns and warming temperatures affect both their roosting sites and the prey upon which they subsist. Vampire bats are climate refugees too. And as they are forced from their traditional grounds, the rabies virus rides along with them.

When I ask Wallace about the degree to which climate change realistically bears on rabies risk from vampire bites, he tells me that the story is, unsurprisingly, a complicated one. "We're closely watching vampire bat rabies that is found in all of South America, Central America, and Mexico," he says. "The vampire-bat home range is theorized to become more favorable in the southern United States—namely, Texas and Florida—over the next fifteen or twenty years, due to climate change." But he is also quick to point out that just because they might soon have a more favorable habitat doesn't mean the bats will necessarily take up residence in the United States. Hence the surveillance apparatus. Wallace's team is tracking vampire bats using everything from citizen science apps like iNaturalist to communiqués from the Mexican public- and animal-health programs. He is concerned enough to be keeping a close eye on the matter.

Vampire bats aren't really keeping Ryan Wallace up at night, though. Unlike the case of primary amoebic meningoencephalitis, we have a vaccine for rabies. Like many vaccines, it doesn't just offer us a shield against the rabies virus; it gives our bodies the tools to forge their own. Wallace has dozens of examples of successful rabies vaccination campaigns.

Instead, his midnight climate worries are all about resource constraints. From his perspective, it's already incredibly difficult to get governments to prioritize rabies. "If we as a rabies community have to compete with food scarcity, natural disasters, increasing political turnover, and other challenges that are all going to be exacerbated by climate change, rabies will fall far, far down on the priority list," he says. "And I think there are certain countries and certain regions where it's going to make it incredibly challenging to see progress."

In other words, Wallace's concerns look more like the Pentagon's than they do those of a climate scientist. It's the manner in which a changing climate acts on social forces that is most concerning. Rabies vaccination programs are largely sponsored by governments. These programs work, he argues: With well-operated programs comes a corresponding drop in rabies incidence. But "if you don't have a functional government," he says, "you're not going to be good at getting money for a neglected disease program. You're not going to be very good at implementing logistically challenging mass vaccination programs."

His logic hints at some of the inequities amplified by a warming world. Part of the great injustice of climate change is that the brain-disease burden will be disproportionately borne by people in the Global South. The disproportionality here is twofold: first, in terms of these patients' contributions to the climate problem—which are often vanishingly small in comparison to those of the people and economies of richer nations—and second, in terms of the political and healthcare infrastructure stress-tested by the diseases in question.

Consider the global vaccination efforts provoked by the coronavirus pandemic. A team of British and Dutch researchers studying a global dataset of vaccine rollout rates noted in 2022 that, within the first year the jabs were available, high-income countries had vaccinated 75–80 percent of their populations, while low-income countries had vaccinated less than 10 percent. The scientists called the apparent disparity in vaccine access "one of the greatest failures of international cooperation during the SARS-CoV-2 pandemic." Such inequities do not bode well for the brain diseases of climate change.

These inequities are also not limited to zoonoses and other infectious diseases: Environmental factors can explain much of the variance in who lives with common neuropsychiatric conditions. While interactions between genetics and the environment aren't particularly helpful in accounting for the incidence of some neuropsychiatric disorders like depression or substance abuse, "they explain a rather large portion of the phenotypic variation of the remaining disorders: over 20 percent for migraine and close to or over 30 percent for anxiety/phobic disorder, attention-deficit/hyperactivity disorder, recurrent headaches, sleep disorders, and post-traumatic stress disorder," wrote a team of University of Chicago researchers in 2022. Environmental factors on their own, irrespective of genetic interactions, can explain additional variance. In practice, that means if you know enough about someone's genetic background, as well as their history of exposures to various environmental stressors—like poor air and water quality—you can estimate with reasonably high confidence their risk of developing a variety of brain ailments. As with the case of cancer's Armitage-Doll theory of the 1950s, these stressors constitute some of the various steps on the stairway to brain disease.

Unsurprisingly, as with climatic factors like extreme heat and extreme weather, these environmental stressors are borne disproportionately by people with low incomes and skin that isn't white. And as Ryan Wallace reminds me, some of the most endemic, highest-risk countries for rabies are already those that sit on the equator.

The inequities in question also compound. The coronavirus, for example, has been linked to a higher risk of neurological and psychiatric conditions two years after infection, relative to other common respiratory infections like the flu. In 2021, one Oxford study reported that a full third of COVID-19 survivors were diagnosed with at least one neurological or psychiatric disorder within six months.

In what has become common parlance in climate circles, it is the feedback loops that offer the gravest diagnoses. According to a literature review led by Cleveland Clinic neurologist Andrew Dhawan, risk factors associated with climate change—like temperature extremes and variability—are associated with everything from stroke incidence to hospitalization of dementia patients to migraine and multiple sclerosis severity. Air pollution, especially nitrates and fine particulate matter, only makes the story worse. (The least pleasant study I've come across in this field has a declarative sentence for a title: "Hallmarks of Alzheimer Disease Are Evolving Relentlessly in Metropolitan Mexico City Infants, Children and Young Adults.")

But there is promise to be found here, in that with deeper understanding of the etiology of brain disease comes the possibility of precision treatment (and prevention) for these disorders. Over the past several decades, scientists have profoundly deepened their knowledge of the human genome. They have begun work on grasping its proteome—the collection of proteins produced by this genome and expressed in particular cells and under particular environmental conditions. You have undoubtedly heard of the microbiome, the body's menagerie of microbes that live on and inside us. Next will be the exposome: a catalogue of the environmental stressors someone experiences over the course of their life—and the manners in which these exposures affect health.

The exposome is largely uncharted territory in medicine, but glimmers of its possibility abound in studies of neuropsychiatric health. Consider the case of schizophrenia. Unlike ALS, schizophrenia is a largely heritable disorder. It is clear from a multitude of twin and family studies that schizophrenia does run in families, suggesting a genetic role in its manifestation. But in the sky of our genome, the disorder isn't quite a constellation, and no single star shines bright enough to be solely responsible. Instead, schizophrenia is more akin to a galaxy composed of numerous smaller stars: hundreds, perhaps thousands of genes, each contributing in its own minuscule way to susceptibility. The heritability of schizophrenia is estimated to be around 80 percent, meaning that about 80 percent of the differences in susceptibility within a population can be attributed to genetic variation. But heritability is not destiny. Environmental risk factors have been shown to explain up to 20 percent of schizophrenia's heritability. Furthermore, in a 2022 statistical study of schizophrenia risk, researchers noted that including environmental and environmental–genetic *interactions* in a model of heritability explained 46 percent of the variance. Heat exposure during early pregnancy is associated with a higher risk of children developing neuropsychiatric conditions like schizophrenia and anorexia, for instance. The exposome matters, and every step toward cataloguing its intricacies is a step toward personalized medicine.

We're not there yet. "Climate change poses many challenges for humanity, some of which are not well-studied," Dhawan said in a press release following his study's 2022 publication on climatic factors and neuropsychiatric health. "For example, our review did not find any articles related to effects on neurologic health from food and water insecurity, yet these are clearly linked to neurologic health and climate change."

Dehydration and a scarcity of nutrients can impair brain development in infants and young children. Iron deficiency, often prevalent in conditions of food insecurity, can lead to cognitive impairment. Under chronic conditions of food and water insecurity, the stress response is consistently activated. The brain, in its vigilance, churns out stress hormones, preparing the body to face imminent threat;

chronic stress, in turn, can change the brain's structure and function, leading to memory problems, mood disorders, and heightened anxiety. As far as I can tell, there is no serious scientific effort yet to understand the neurological relationship between climate change and chronic stress. If we want a shot at warding off the coming neurological nightmare, we have to walk into this thing with open eyes.

As Sandra Gompf, Philip's mother, told the *Tampa Bay Times* shortly after founding the Amoeba Season campaign, "It's important that people enjoy recreation. Philip loved the outdoors. Nature is not evil."

Nature is only ever just nature. It pushes and pulls; sometimes it is in friction with us. Sometimes it displaces. But never does the weight of nature act with normative menace. We have to meet it with the same practicality if we want a healthier relationship with its forces. If you're planning on jumping in a warm lake during amoeba season, plug your nose. Wear a noseclip like you'd wear a seat belt in a car. Maybe it sounds a little funny—a little overreactive. Maybe it sounds like wearing an N95 to the grocery store three years into a pandemic. But those work too.

It's true, though, that just like pandemic fatigue is real for many people, *Naegleria fowleri* infections are rare—and will continue to remain relatively rare, even under our coming climate futures. Psychologically, it can be difficult to motivate oneself to take these kinds of risks seriously. But consider an infection like Lyme disease, for example, which affects about a half million people in the United States alone every year. As tick habitats broaden, the risk of Lyme's accompanying neuroborreliosis sharpens steeply. And as the brain diseases of climate change march forth, a noseclip might prevent the rarest among them, but it won't stop the ticks and mosquitoes. I know doomsday alarmism is tiresome. But you should still be concerned.

Public health practitioners are in a difficult position. Their field is one of necessarily global ambitions, but its policies play out at the individual level. Preventing zoonotic spillover events is particularly tricky, because it relies on an understanding of a vast web of interactions between people and ecosystems. But it is not impossible.

Let me climb the public health pulpit for a moment. Heightened monitoring of animal populations, especially those known to harbor zoonotic pathogens—as Ryan Wallace's team at the CDC has implemented in the case of vampire bats—can help offer insights into emerging threats. Equipped with this surveillance knowledge, we can implement early warning systems, swift response protocols, and targeted interventions to contain potential spillover events. In safeguarding biodiversity and protecting natural habitats, too, the delicate balance of nonhuman species can flourish undisturbed. Implementing sustainable land-use practices, curbing deforestation, and reevaluating exploitative interactions with wildlife are all part of the story.

They're not the only part. We also need to embrace a paradigm of interdisciplinary collaboration. We need to pay attention to the manners in which human and animal health are already woven together. In bridging the gaps between medical professionals, veterinarians, ecologists, and local communities, we can forge a stronger, holistic front against zoonotic threats. People need the tools and resources to detect, prevent, and respond to potential outbreaks. We need education campaigns to help illuminate the path toward prevention. In disseminating accurate information, public health officials and governments can cultivate an understanding of zoonotic risks and help foster responsible behaviors. Empowering people to make informed choices about their interactions with animals, advocating for proper hygiene practices, and promoting the responsible use of antibiotics are all necessary tools in the toolbox of spillover prevention.

It all sounds great, right?

Public health pulpit relinquished. We need this spirit and these interventions, but many of them are lofty goals. They are difficult to obtain and implement. And as Wallace suggests, public health infrastructure (and funding) often relies on political stability, the likes of which can also be threatened by a changing climate. In the absence of a well-coordinated global response to spillover risk—and it needs to be global, because it's not like these critters are paying attention to political borders—we're left with one another: with mutuality. And if the coronavirus has taught us anything, it's that partisan politics can trump science, and remarkable swathes of the populace (perhaps you and me included!) can become desensitized to mass death. These facts do not inspire confidence in the face of slow-moving catastrophes like climate change and shifting animal habitats.

But mutuality does have an intimate place in public health policy and practice. It's not nothing. We know our collective well-being is intricately intertwined. There's a call to arms here: to acknowledge the profound interconnectedness that binds us as individuals and communities, and to approach public health challenges with a spirit of shared responsibility and solidarity.

What might that look like in practice? I think it has something to do with implementing policies that acknowledge that the health of each thread in the tapestry of life—each individual—impacts the integrity of the entire weaving. Such policies recognize that the warp and weft of this grand design are inseparable, that an illness in one thread can unravel the whole, and that the strength of each thread lends resilience to all. Consider, for instance, public health policies aimed at addressing health disparities. Such efforts are rooted in the notion that when one group suffers, it diminishes the health of the whole community. Providing resources to those most vulnerable, therefore, is not an act of charity but an investment in collective well-being. From sanitation infrastructure to mental health initiatives, from food safety regulations to programs addressing homelessness, mutualistic public health policy is a living testament to our interconnectedness. It is a reminder that we are not isolated entities but part of a vast, dynamic organism.

A shining example of this approach comes from the case of harm reduction, the set of practices referring to compassionately and directly engaging with people who use drugs in the interest of preventing overdoses and the transmission of infectious diseases. In the realm of substance use, harm-reduction interventions have demonstrated stunning results. Needle- and syringe-exchange programs, a cornerstone of such efforts, have been shown to drastically reduce the transmission of blood-borne infections like HIV and hepatitis C among people who inject drugs. Harm-reduction programs not only provide access to sterile equipment but also serve as points of connection to healthcare and support services. The distribution of naloxone, too, an opioid-overdose reversal medication, illustrates further the promise of the approach. By equipping people, including friends, family, and community members, with the medication, harm-reduction initiatives have helped prevent countless deaths from opioid overdoses.

There is a general truth to be realized here about realism. The brain diseases of climate change are coming, but we can meet them head on. Pretending they don't exist—as in the case of, say, municipal policies that sweep homeless encampments out of sight without implementing permanent housing solutions—isn't going to help. Harm reduction is a philosophy that invites us to meet nature and ourselves where we are, not where we wish we could be. It is a doctrine that acknowledges the imperfections of the world—and, instead of imposing harsh, often impractical prohibitions, strives to minimize the risks and harms associated with those imperfections. Zoonotic diseases are as old as our interactions with the wild. Rather than attempting to erect impermeable barriers between humans and animals, we can focus on reducing the opportunities for disease transmission. And we can come to one another's aid when tragedy strikes.

This is where education—already noted as a cornerstone of effective public health policy—reenters the picture. Education can empower us to understand and respect the boundaries between humans and animals. It teaches us the significance of biosecurity, of safe food practices, of sustainable land and water use. In doing so, it helps erect a buffer—not a barrier—between people and zoonotic diseases. And it plays a pivotal role in shaping our collective response. By definition, education helps foster shared understanding, a united front against the spread of the brain diseases of climate change. It is not merely a tool or strategy. It is the sunlight that allows us to see the coming wave.

In July 2022, another boy, a few years older than Philip Gompf, began complaining of a headache after swimming in brackish water south of Tampa. It was warm. The headaches intensified, and neck stiffness followed. In fact, Caleb Ziegelbauer's story sounds similar to Philip's in all ways but one: At the time of writing, Caleb is alive. Doctors quickly recognized what had happened, and Caleb received the necessary drug cocktail. He may possess some underlying immunity. We don't know. We do know he is lucky.

In Caleb's case, that doctors understood the potential problem here at all—that's education working. That is the Amoeba Season campaign in action.

How can we expect doctors to recognize threats they may not have learned about in medical school? One answer lies in understanding that a physician's education is never static, but continually reshaped by the forces of inquiry, observation, and lifelong learning. A doctor's education doesn't end with school. It blends with the experiences gathered over a lifetime, and it is enriched by continuous investigation. As new infectious brain diseases emerge under a changing climate, physicians have a fleet of opportunities at their disposal. Scientific journals, professional conferences, and continuing medical education courses can guide them through the rapids of new information. Interactions with colleagues and advocacy groups like the Gompfs', discussions with experts in infectious disease, and participation in professional networks serve to pool collective wisdom, making it easier to recognize and respond to new threats.

Amid this torrent of knowledge, an important skill for a physician is to remain intensely patient and observant. Recognizing the signs of an emerging infectious disease often involves paying close attention to the subtle changes in the clinical landscape, picking up on patterns that may hint at a new threat, and maintaining an open, questioning mind. In this manner, when new threats emerge, they can be met with the combined might of accumulated wisdom, vigilant surveillance, and an ever-deepening understanding of the vast, interconnected landscape of human health.

What about the rest of us? Viruses and bacteria are constantly evolving; the climate continues to change. We shouldn't expect one another to remain as abreast of developments in infectious disease as we might expect of a doctor. But we do need to listen to health authorities. Perhaps the CDC's credibility suffered during the coronavirus pandemic, but it and the World Health Organization remain the watchtowers in the landscape of disease emergence, offering early warnings and insights from the front lines of research and observation. We can stay curious and informed. We can listen to the storytellers here, the scientists and researchers who dedicate their lives to decoding the madness unfolding before us. And we can invest, ourselves, in mutuality. I suppose even something like a GoFundMe page offers an outpost for this mode of solidarity and investment in one another.

Caleb Ziegelbauer's is called "Wake up, Caleb." In October 2022, three months after his infection, his aunt Katie posted: "IT'S DECANNULATION DAY!!! CALEB'S TRACH IS OUT." Slow progress, but real progress nonetheless. In the coming months, he'd learn to sit up, to swing a Wii baseball bat. By March, he'd be home. Plenty of intensive therapy left, to be sure—he has a long road ahead. But Caleb is, in fact, awake now.

Are we?

6

THE BODY
KEEPS THE STORM

it starts

                                        with smoke

        it always starts   with smoke
    —Jazz Money, "sweet smoke," 2020

Pain.
    —Toni Morrison, *Jazz*, 1992

There was a fire in Gatlinburg. "There was a fire in Gatlinburg": I wish that sentence were the only one in this story. I mean: I wish the fire were the end. For Michael Reed, it was the beginning.

When Constance called Michael on November 28, 2016, he could hear a wavering concern in his wife's voice. There had been a fire outside of town, so they'd all understood, and now she could see flames coming from a house across the street. Michael and their son Nicholas had taken the family's only car into downtown Gatlinburg to try to get more information about what was happening. On the phone, separated from Constance and their two daughters, he told her to call 911.

Moments later, on the phone with the dispatcher, she says: "The fire is next door to my house. . . . My husband is not home. I don't have a vehicle, and I have no way out of here. I have no way out, and I have children at home." There is some hurried discussion. The dispatcher tells Constance: "Stay with me." The line goes dead.

Michael didn't hear this call. He couldn't have; he was downtown. A park ranger told him to go to the nearby resort town of Pigeon Forge. Presumably, eventually, he and Nicholas would reunite with the rest of their family there. They left.

As autumn had descended on the Great Smoky Mountains that year, an unseasonably hot, drought-ridden stretch of time and space had turned this normally lush and verdant expanse of Appalachia into a kindling bed. In November, the requisite spark arrived, ultimately birthing one of the most devastating wildfires the region has ever witnessed. It had begun innocuously enough, a silent flicker amid the vast wilderness of the national park. Two teens, their intentions as yet undetermined, were later charged with starting the fire by tossing lit matches onto the parched ground near Chimney Tops Trail. The blaze began to feed and grow, undetected at first, then underestimated.

Flames would gallop toward Gatlinburg and the Reeds, swallowing homes and histories in their wake. The inferno painted a grim tableau on the land. Buildings that had stood as silent custodians of human lives and memories became skeletal remains. Vehicles lay abandoned, their shapes distorted by oppressive heat. The air, once crisp with the scent of firs, was choked with smoke and an acrid tang.

Michael had found himself in an unimaginable scenario. He wanted to brave this smoking chaos and searing wind and ember rain. He wanted to find the rest of his family. But he couldn't get home. The fire simply didn't allow it.

He would search the wreckage for days. In the aftermath, after nearly a week, the news arrived that Constance, Chloe, and Lily had not survived the wildfire. Their bodies were found in a nearby house.

Much of these events I have reconstructed from news reports and court filings. When I talk to Michael Reed, we don't talk much about Gatlinburg, or about Constance or Chloe or Lily. Instead we talk about his federal tort claim against the United States—a claim that argues for damages arising from the "negligence of employees of the National Park Service." Namely, he sued the country because he couldn't sue the drought, the likes of which in Tennessee were more or less historically unprecedented.

Frankly, though, we spend most of our time talking about the case because I do not know how to write about a man losing his life partner and two children in a drought-fueled wildfire, other than to write that it happened. This is the land of no simile. They died in the fire like people who die in fire.

Simile and metaphor work for stress, though. Imagine a lone deer grazing in a meadow. The slightest rustle in the underbrush—the faintest whiff of an unfamiliar scent—and the deer's senses sharpen. Its muscles tense, ready to bolt. This reaction constitutes a normal stress response: an essential survival mechanism finely honed by millions of years of evolution. Like the deer, our bodies, too, react to perceived threats with a heightened state of alertness. The heart pounds, the breath quickens, the senses stretch, taut. This acute stress reaction, the fight-or-flight response, is an integral part of our shared heritage with the natural world. It readies us to face danger or to flee from it, marshaling our resources for survival. And once the threat has passed, just as the deer gradually relaxes when the rustle in the underbrush proves to be only the wind, our bodies return to their baseline state.

In the aftermath of a deeply traumatic event, the natural rhythm of the stress response can deteriorate. Post-traumatic stress disorder (PTSD) is a condition well captured by its name. It refers to something bleakly specific and medically well-defined, and I deploy the phrase in that sense. PTSD has had a home in psychiatry's bible, the *Diagnostic and Statistical Manual of Mental Disorders (DSM-5)*, since its third edition was published, in 1980. Here, far from a transient rustle in the brush, the threat is a wildfire that has left deep scars in the landscape of the mind.

Intrusive thoughts are perhaps the most palpable of these scars. Vivid memories of the traumatic event can persist, punctuating everyday calm with recurrent flashbacks and causing sufferers of PTSD to relive the trauma over and over. Nightmares are frequent visitors in the quiet hours of night. Avoidance of places, activities, or people that remind those with PTSD of the trauma—an attempt to circumvent these intrusive thoughts and feelings—can offer another notable symptom. The behavior can lead to a gradual withdrawal from the world.

Yet the condition is *also* characterized by hyperarousal, which can keep people in a state of constant alertness. In practice, hyperarousal might manifest as difficulty sleeping, concentration problems, a quick temper, or an exaggerated startle response. Through this lens, life appears fraught with unseen danger. PTSD's symptomatology can combine to spur broad, negative changes in cognition and mood. It completely changes one's experience with the world. Feelings of guilt, self-blame, or a lingering sadness can pervade day-to-day existence. Once-pleasurable activities might lose their appeal, and a sense of emotional numbness can set in.

Neuroscience is a useful crutch for understanding some of this dislocation. In the brain, we can zoom in on three areas particularly relevant to post-traumatic stress. Consider, first, the amygdala, an almond-shaped neural sentinel of sorts. One of the area's primary roles is to alert us to danger; the amygdala activates the fight-or-flight response when the shadow of a threat looms. In the wake of severe trauma, the area can become overreactive, firing off alarms at the slightest hint of danger—or even at innocuous stimuli that merely resemble the original source of trauma.

The hippocampus, too, a seat of memory in the brain, bears the scars of PTSD. Like a tree charred by lightning, the hippocampus can shrink under the corrosive influence of chronic stress hormones. This atrophy can lead to memory instability and deficiencies in the brain's ability to discriminate between past and present experiences. The catalyzing traumatic event, rather than being filed away as a past occurrence, can feel ever present. In essence, PTSD affects the hippocampus's ability to contextualize memories, resulting in flashbacks that are experienced as if they are happening today.

Finally in the neuroscience of trauma, there is the prefrontal cortex, the brain's maestro. In a healthy body, this region helps regulate our emotional responses, quieting the amygdala's alarm when danger has passed. However, in people with PTSD, the prefrontal cortex can fail to perform this role, allowing the amygdala to blare uninterrupted—and leaving the person housing this brain with prolonged periods of anxiety and hyperarousal. In its interactions with memory systems, the region can struggle to quell the repeated outbursts of the hippocampus.

The neurological story allows us to orient ourselves with a model of how trauma unfolds within us. Yet despite our reasonable grasp of the underlying neuroanatomy here, no brain surgery will cure post-traumatic stress. In part, that's because trauma isn't limited to our heads. Trauma is an echo reverberating deep within our bodies. It pulls strings and flips switches, disrupting the finely tuned mechanisms that maintain equilibrium. Among people with PTSD, the autonomic nervous system—the body's control panel for fight-or-flight responses—is persistently on high alert, pushing the body into a state of constant arousal. The result is akin to flooring a car's accelerator and expecting the engine to function normally. The impact is systemic and broad.

The cardiovascular system is a prime target. Hearts under the influence of PTSD beat a more erratic rhythm, working overtime and straining against a tide of stress hormones like adrenaline and cortisol. Blood pressure can be persistently high, increasing the risk of heart disease and stroke. Digestion, too, bears the brunt of this relentless barrage. Under stress, our bodies divert resources from "nonessential' functions like digestion to cater to the perceived threat. Those with PTSD may wrestle with bouts of nausea, loss of appetite, stomach pain, or irritable bowel syndrome. The gut, home to an entire ecosystem of microbes, is thrown off-balance, which can further degrade physical and mental health.

The immune system isn't immune to PTSD's effects, either. Chronic stress can weaken the body's defenses, making people with PTSD more susceptible to infections and illnesses. They may find that they catch colds more easily or struggle to shake off bugs.

Prolonged stress can accelerate aging at the cellular level, shortening telomeres, the protective caps at the ends of our chromosomes. Nightmares and insomnia can rob the body of restful sleep, which it needs for repair and restoration. This sleep deprivation is a stressor in itself, escalating the cycle of physical strain and further fueling the body's chronic state of alertness.

Exposure to trauma is common. Assaults, accidents, other forms of violence—researchers estimate that around half of adults in the United States will experience some form of violent traumatic event over the course of their lives. But most don't develop PTSD: The lifetime prevalence is closer to 7 percent. Susceptibility remains something of a mystery.

What isn't a mystery is that you don't need to go to war to experience the disorder. In the quiet drama of everyday life, far removed from the din of battlefields and the specter of military conflict, trauma can stay with us all. Sufferers of PTSD are survivors of accidents and natural disasters, victims of personal violence or abuse, witnesses to horrific events, or carriers of deep emotional wounds from childhood. Each story is as unique as a fingerprint, yet all share a common thread: the experience of a trauma that lingers, resurfaces, and shapes the course of a life.

When we first connect, Michael Reed tells me that he suffers from PTSD spurred by the trauma of that night in November 2016. It came quickly and harshly. Colloquially, we often understand PTSD as a condition of combat veterans. His is a reminder that there is great trauma to be found at home. "To say I'm broken is an understatement," he says.

For Reed, the brokenness began as a purposelessness. He had become unmoored. He couldn't think straight. *What was it to live if not with his family?* he asks.

In the late days of October 2012, a tropical cyclone merged with a chilling northern front in the Atlantic. Given the name Sandy, the superstorm was of a scale that nearly defies comprehension. It approached the coastline like a hammer. In New Jersey and New York, the storm surge was like nothing before seen, more akin to a wall of water than a series of waves. It invaded streets, transformed highways into rivers, and swallowed homes whole. The boundaries between land and sea were obliterated. In New York, the city of steel and glass—the city that never sleeps—Sandy plunged neighborhoods into a darkness punctuated only by the eerie glow of flooded substations. The subway system was choked with brackish, corrosive brine. Aboveground, the wind seized trees and power lines alike, leaving in its wake a landscape strange and unrecognizable.

As a changing climate makes natural disasters more extreme and more frequent, we should expect to see the occurrence of PTSD rise in kind. The logic is similar to that of the brain diseases of climate change: Even if the underlying human susceptibility to the condition remains constant, the climate-fueled rise in risk factors will increase, and cases will rise accordingly.

"We're all on a first-name basis with the climate crisis. Sandy, Maria, Irma, Dorian—my fellow plaintiffs and I are directly impacted every day." Vic Barrett was speaking at a 2019 press conference splayed out in front of the Supreme Court. Then a twenty-year-old from White Plains, New York, Vic is suing the federal government over its inaction on climate change. The hurricanes he names are the realities of the untamed environment in which he grew up: When Sandy hit New York in 2012, Vic's home lost power, and his school year came to a grinding halt.

Today, Vic's doing okay, but others in the path of hurricanes can't say as much. In 2010, for example, researchers studying the impact of Hurricane Katrina on the mental and physical health of low-income parents in New Orleans found that about half of the nearly four hundred study participants were most likely suffering from PTSD. The effects were directly related to the storm: The worse your experience of the hurricane—in terms of exposure and property damage—the more likely you were to experience symptoms of mental illness, PTSD, and perceived stress.

Yoko Nomura's study is even more stunning.

A psychologist in New York City at the CUNY Graduate Center and Queens College, at the time of Sandy, Nomura had already assembled a cohort of expecting mothers in preparation for a study on stress and birth outcomes. She had intended to measure the effects of differential stressors on mothers and children, teasing out the potential contributions of various forms of environmental stress to psychiatric conditions later in life. And then one of the greatest environmental stressors of all waltzed into town. Sandy left pregnant women trapped in elevators and without access to water. It upended life.

Nomura's resulting study almost beggars belief. What she showed in late 2022, in a ten-year look back on Superstorm Sandy, is that children in Sandy's path who were in utero during the storm now carry a severely disproportionate risk of psychiatric conditions, relative to kids born before the storm or who were conceived after it. Compared to girls who were spared from the storm, for example, girls who experienced Sandy in utero saw a twenty-fold increase in anxiety and a thirty-fold increase in depression. Boys saw a sixty-fold increased risk of ADHD and a twenty-fold increase in conduct disorder. Symptoms presented as early as preschool. Even in the ascetic prose of an academic study, Nomura and her colleagues called their findings "extremely alarming." As she told *The Washington Post*: "I did not expect this to be so clear-cut."

The study rings the alarm bells of climate change. It also offers insight into the notion of intergenerational trauma. In recent years, science has taken us deep into the genetics of trauma, unveiling how such experiences might find a path from one generation to the next. Intergenerational trauma is not merely a metaphorical hand-me-down: It is a tangible inheritance, carried on the double helix of our DNA. Epigenetics, a relatively new frontier of science, refers to changes in gene expression induced by environmental circumstances that don't necessarily involve alterations to the underlying DNA sequence. Traumatic experiences, it appears, can lead to such modifications—and intriguingly, these changes might be passed to offspring.

Studies like Nomura's, and further, those observing populations deeply scarred by catastrophic events—Indigenous communities ripped apart by colonization, survivors of the Rwandan genocide, descendants of those who survived the trauma of slavery—reveal that the children and grandchildren of people exposed to severe trauma often struggle with anxiety, depression, PTSD, and other psychological challenges at higher rates than the general population. In one landmark study, mice that had learned to associate a particular smell with a mild shock produced offspring that were also anxious about this smell, despite never having encountered it before. Meanwhile, human research has shown that children of people who survived the Holocaust have altered stress hormone profiles, despite not having experienced such extreme trauma themselves. Trauma is not a thing you're born with, unless it is.

On a trip to Two Dot, Montana, I drive through five hundred miles of smoke. It is early September, and forty wildfires have burned in Idaho and Montana in the past month alone. The largest, the Moose Fire in the Salmon-Challis National Forest, crosses the hundred-thousand-acre mark the day I inch eastward across Idaho. Southeast of Missoula, the air quality index is nearly three hundred: the point at which the government issues health warnings indicating emergency conditions. Along some sections of Highway 294, lurching hills induce a kind of nausea: The smoke blots out any semblance of a trusty horizon, and a cresting road seems to suggest I'm about to tip over the edge of the world. I pull over and look back. As it sets, the sun is a golden nugget; the sky, muddied, the color of a river silt-stained by panners. What are we looking for?

I recall an editor telling me that she couldn't attend backyard barbecues without being set on edge. She lives in California, and one whiff of charcoal would bring back the horrors of wildfire season.

Looking out through the roadside haze, thinking of this editor, Teresa, I'm reminded that our brains are designed to be associative. Occasionally, and for good reason, they associate sensory inputs with danger. This primal system of association is meant to keep us safe, imprinting our neural pathways with patterns that alert us to potential threats. It's when trauma comes into play that this system of associations can turn from a protective mechanism into a maze of unexpected triggers. A memory of trauma isn't like a neatly catalogued file in an archive; it's more like a knot of roots under the forest floor. Seemingly unrelated stimuli—a certain smell, a particular shade of twilight, a tone of voice—can touch upon these roots and send a shock of recollection surging through the system. These associations can even alter the landscape of memory, infusing innocuous elements of the present with echoes of the past. A benign thunderclap becomes an incoming artillery shell; the flicker of a car's headlights revives the glare of an accident; the gentle pressure of a handshake becomes the crushing grip of an assailant.

Is it any surprise these associations bear on our behavior? Trauma is a coiled force, and it makes us act. Certainly, sometimes it makes us *stop*—we freeze—but active prevention of behavior is just another type of action. Like temperature-fueled aggression, then, environmental trauma is another thief of agency. Post-traumatic stress steals our decisions. Can anyone really claim they *want* to drink to forget? That they *want* to freeze up when they hear the wind? This is the environment reaching in and tipping our scales, making decisions without our conscious consent. It's how we wind up as puppets, with the climate as our puppeteer.

The physicality of environmental trauma shouldn't surprise us. From the perspective of embodied cognition—of a brain resting in a body; of a mind only making sense in the context of that body—trauma is bigger than the organ with which it's most closely associated. Our internal signal cascades, those tumbling ebbs and flows of neurotransmitters and hormones and other biological signaling molecules: They permeate the body; they mirror a violent, crashing landscape. There is a reason Stephanie Foo's book on complex PTSD is called *What My Bones Know*. There is a reason Bessel van der Kolk's book on trauma is called *The Body Keeps the Score*. The brain is just the referee.

Given post-traumatic stress's whole-body effects, some researchers consider it to constitute an inappropriate ability to extinguish fear—and in particular, to extinguish a kind of aberrant, physicalized, associative fear. You can train a rat, for example, to fear a tone: If you trap it in a cage and deliver a little electric shock to its foot every time you play a sound, the animal will eventually begin to freeze up when it hears it again. That's basically just Pavlov. But scientists can use this freezing—the conditioned fear response, as it's known in neuroscience and psychology—to measure a rat's ability to *forget* the association in question. A rat doesn't naturally fear a tone. Remove the associated foot shocks, and eventually they'll stop freezing when they hear it. You're already familiar with the neuroscience of active forgetting.

All animals are somewhat like this. We learn about the world—we model the relationships encoded in it—and then we update our beliefs based on new sensations impressed upon our brains. It may be unnatural to associate a tone with a shock, or the sound of wind with grief, but with dramatic evidence, our models update accordingly. In some sense, our brains are doing exactly what they're supposed to be doing. We *ought* to be stressed post-trauma. Trauma is stressful. But post-traumatic stress becomes a disorder when we can't unlearn the association between the tone and the shock.

Framed as a problem of fear extinction, though, PTSD lends itself to a variety of behavioral treatments. Through this lens, the broad goal is to facilitate a natural recovery process—to encourage new growth around the scar of trauma and allow for fear to be remembered but not relived. In a therapeutic context, for example, this process might take the form of exposure therapy, where people are gently and systematically exposed to reminders of the traumatic event within a safe and controlled environment. Much like how controlled burns can facilitate a forest's healing process and make it resilient to future fires, this approach seeks to stoke fear in manageable doses, allowing people to experience it in a new context: one of safety rather than danger.

Similarly, therapies such as cognitive behavioral therapy (CBT) and eye movement desensitization and reprocessing (EMDR) work to shift the relationship between the individual and their trauma, moving unprocessed memories to a place of benign, but still remembered, safety. CBT is a journey into thought patterns and behaviors— a careful sifting through the layers of the mind. It operates on the belief that our thoughts, rather than external events themselves, determine our feelings and behaviors. In some sense, it's akin to a mental courtroom, in which irrational fears and thoughts are put on trial, and their credibility and influence are scrutinized under the clinical gaze of the therapist and patient. Trauma can skew our perception, painting a distorted picture of reality that maintains the cycle of distress. CBT seeks to challenge this picture: to redraw the lines with a more rational and balanced perspective.

EMDR takes a different path. Developed in the late 1980s by psychologist Francine Shapiro, the approach combines elements of various therapeutic practices with a unique ingredient: bilateral stimulation of the brain through guided eye movements. In an EMDR session, the patient is asked to recall traumatic events while receiving some type of bilateral sensory input (such as side-to-side eye movements or hand tapping). It's as if the therapist and patient are working together to reroute a memory, adjusting its emotional charge and the way it is stored in the brain. The exact mechanisms of how EMDR works are still under investigation. Some theorize that the process is akin to the rapid-eye-movement stage of sleep, which plays a role in how memories are stored and processed. Others suggest that the dual attention—focusing on the traumatic memory and the sensory input simultaneously—can help change how the memory is held in the mind.

These therapies don't erase the past, but they offer the tools to understand and reframe it in a manner such that trauma no longer holds the same power over the present. They help ensure that the natural processes of fear extinction can be facilitated, encouraged, and coaxed into life again.

On November 25, 2018, the California Department of Forestry and Fire Protection announced that it had successfully contained the Camp Fire. But the containment did not spell the end of the madness. At the time of writing, there are still bodies to be identified. The Camp Fire is the deadliest fire in California history.

It had taken its name from its place of origin, Camp Creek Road, but as the days unfolded, the connotation of "camp"—a temporary dwelling, a brief human intrusion into the wild—grew painfully ironic. There was nothing fleeting about the impact of the blaze. The fire spread with a terrifying ferocity across the northern expanses of the state. The town of Paradise bore the brunt of the inferno; it was nearly swallowed whole. The fire created its own weather: Towering, so-called pyrocumulus clouds rose skyward, a testament to the heat and energy of the blaze. Over the course of seventeen days, the Camp Fire ravaged more than 153,000 acres, reducing homes, forests, and memories to ash. Almost 19,000 structures burned; 30,000 people lost their homes.

Jyoti Mishra is a cognitive neuroscientist and director of UC San Diego's Neural Engineering and Translation Labs. In the immediate aftermath of the fire, Mishra began surveying survivors of the nearby town of Paradise for evidence of post-traumatic stress. She found that people who had been indirectly exposed to the fire—perhaps they had a family member in its path—had a rate of PTSD roughly three times that of the general population. People who were directly exposed, who had actually lost property, for example, had a rate three times higher still.

Mishra, who also codirects the Climate Change and Mental Health Initiative at the UCSF Center for Climate, Health, and Equity, grew up on a hospital campus. Her parents were doctors. She saw mental health needs all around her; she notes a history of mental health concerns in her family as well. In some important sense, she was somewhat destined for work in medicine.

Mishra is interested in grasping a causal model of trauma. She wants to understand exactly what it is, where it comes from, and what it does to us. And by extension, she wants to understand how we can best respond to it.

In the 2010s, after finishing her graduate studies, she began working with international organizations on childhood trauma. It was then that her work took a curious turn. Mishra had been studying foster children—in particular, those with higher incidences of ADHD. As a translational researcher, she'd been trying to understand what kinds of interventions might be most effectively implemented to help these kids with distractibility. One of the tools in question was based in cultivating a practice of mindfulness: a set of trainings focused on self-regulation and establishing a mindful connection with the environment.

Generally speaking, mindfulness refers to a kind of attentive and nonjudgmental presence. To experience the world mindfully is to stand on the shore of our own experience and bear a kind of quiet witness to our inner world—the eddies of thought, the surges of emotion, the glistening pebbles of sensory detail—without necessarily trying to change the direction of the flow. In the clinical halls of academia, it can be a little hard to pin down. But for Mishra and her colleagues studying ADHD, it was early days yet. The researchers were throwing a lot at the wall, and they wanted to know what stuck.

Surprisingly, all else equal, it was the mindfulness tools that consistently showed the largest benefit for the children's distractibility. Mishra couldn't spin it any other way. It was via these initial studies that mindfulness began to creep its way into Mishra's translational work more broadly. Today, it's a prominent feature. And it could have much to say for survivors of natural disasters too.

Imagine PTSD as a thicket—tangled, impenetrable—in the forest of our minds. Researchers like Mishra are beginning to understand mindfulness as a kind of gardener. It does not clear-cut or burn away the thicket; instead, it slowly and patiently begins to explore the rough terrain, acknowledging the reality of its existence and accepting its presence within the greater woodland. Just as you can't force a wild creature to come out from its hidden lair, the memories and echoes of trauma will not be coerced into healing. Like an observer in the wild, sitting quietly until animals grow accustomed to her presence, so does the mindful practitioner cultivate a space of safety and nonjudgment in which these wounded parts of self can come forward, be seen, and be acknowledged. Mindfulness invites a sense of connectedness with the present moment, a sensation of being firmly rooted in the now.

PTSD, as we've seen, often involves a kind of temporal dislocation. Past traumas break through into the present, overwhelming our sense of safety and control. By grounding us in the ongoing flow of our sensory experiences, mindfulness can help us make our way through the darkness. The wilderness of the mind, like any wild space, can be daunting and fraught with danger and uncertainty. But it is also a place of extraordinary beauty, resilience, and adaptation.

Through mindfulness, so goes the thinking, we can come to see PTSD not as a blight upon our landscape but as part of the ecosystem of our selves. We can learn to tend to our scars with compassion, to weather our inner storms with grace, and to reclaim our capacity for growth and transformation. We can learn to make peace with our traumas, incorporating them into the greater narrative of our lives.

In a 2023 follow-up study of Camp Fire survivors, Mishra realized something important about her study participants: Namely, they looked like her distractible foster children from a decade previously. Mishra and her team tested subjects' cognitive function across a range of factors, including attention, memory, and impulsiveness, to name a few. They wanted to understand how climate trauma, in particular, affected the brain. They also measured participants' brain function while they performed a variety of cognitive tasks, using recordings obtained from electroencephalography. What they found was that both groups of people exposed to the fire (the same "direct" and "indirect" groups cited earlier) dealt specifically with distractions less easily than people who weren't.

The findings made sense to Mishra. "A threatening environment tends to make you hypervigilant," she tells me. "Everything in your environment can be a relevant signal." And when you're constantly aware, you're readily distractible.

"And this was by chance," she says, referring to the distractibility finding. "It's not like we knew interference processing would be deficient here." Now, though, her studies on mindfulness feel all the more relevant. The researchers haven't yet explored the application of their mindfulness tool kits—much of which focus on short, five- to ten-minute bursts of breath work—specifically to climate trauma. "But in other work related to trauma," she says, "we've shown that that kind of deficit can go away when one does a full mindfulness-training type schedule. So we wonder what that kind of dissemination would do in this community."

Earlier, we encountered the work of Nadia Gaoua, who studied the relationship between heat and cognition. Part of her implicit claim was that test scores plummeted in the heat because of heat's discomfort, which we find distracting. If she's right, and our cognitive constraints have more to do with discomfort and distraction—the juggling act—than any deep thermal change in the brain, it would seem the neurological antidote to heat's poison would come by way of presence and acceptance, as well. One of the interesting things about temperature spikes, as studied in the lab, is that they decrease our reaction time on cognitive tasks. Our accuracy on a given problem set drops, but we're often performing more quickly. It's like we've become more impulsive. Heat, then, in Gaoua's contemporary understanding, looks behaviorally a lot like distractibility in terms of its ability to sap cognitive and attentional resources from working memory. And that's actually good news, because researchers like Mishra are showing us that we don't need to remain stuck in distraction.

Anna Jane Joyner's passion for environmental justice didn't arise in a vacuum. The daughter of a prominent evangelical pastor in Alabama, she was raised immersed in a world where faith was a living, breathing entity—a guiding principle that shaped one's worldview and relationship to others and to the divine.

But within her grew a keen awareness of another divine relationship: that between humanity and the Earth. Today, as a climate activist, she often speaks the language of her faith to bridge the chasm between the evangelical community and environmentalism. She intertwines biblical teachings with the call for climate justice, urging fellow Christians to recognize caring for the Earth as a moral and spiritual responsibility. Her work has spanned numerous arenas, from cohosting the podcast *No Place Like Home*—which explores the spiritual and cultural dimensions of climate change—to appearing in the documentary series *Years of Living Dangerously*, where she attempted to encourage climate-change conversation within the evangelical community.

These days, Joyner mostly thinks about storytelling. "Humans started telling stories roughly seventy thousand years ago to navigate the harder parts of life: to navigate death and loss and all these difficult parts of being a human," she tells me. "And so it just seems like we absolutely need stories to be able to navigate living in a world of climate change." There is something in writing through trauma—or, at least, in its narrated expression to others. It's a theme I hear again and again in these types of conversations: the value of storytelling as a mode of navigating climate trauma and PTSD.

It makes sense. When we are deeply immersed in a story, our brains exhibit a phenomenon known as neural coupling. The idea is akin to a harmony—and is a profound kind of empathy—a neural echo that blurs the line between the teller and the listener. Functional magnetic resonance imaging studies have offered intriguing insights into the process. Researchers have found that not only does the listener's brain activity mirror that of the speaker, for example, but it can also predict it. The stronger the neural coupling between the two individuals, the better the listener understands the story. In essence, neural coupling is a manifestation of shared understanding, a mental meeting point where two minds, for a moment, converge.

But neural coupling doesn't just enhance understanding: It also helps us anticipate and respond to others. It plays out in the subtle cues of conversation, in the timely nod of agreement or the mirrored laughter. It's there when a yawn ripples contagiously through a room or when a group of friends start to finish one another's sentences. It's in the inexplicable sense of connection we feel with certain people.

In other words, storytelling is more than a mere act of communication. It is a complex, dynamic process that engages our brains in ways that are as rich and varied as the patterns one might find in a bird's feather. In Joyner's telling, stories reinforce our interconnectedness, our shared humanity, and our capacity to transcend the boundaries of our individual experiences. They remind us that we are all, in our own ways, both the story and the storyteller.

Consider writing. When we write about trauma, we necessarily become both the teller and the listener. We have to read the words on the page. We have to traverse the territories of our own pain. The act of storytelling allows us to mold our afflictions into words. It's not merely about recording experiences; it's about transmuting our suffering into something graspable, something we can hold up to the sunlight of understanding.

In the case of PTSD, then, storytelling can be an act of reclamation. Much of exposure therapy and cognitive behavioral therapy is predicated on this idea. In storytelling, we can wear down the roughness of trauma into softer, more manageable shapes. Writing transforms. It is a metamorphosis, turning the indigestible fragments of experience into something one can begin to comprehend. It invites a certain distance, a perspective that enables the observer to consider the situation from outside the immediacy of the event. By giving form to what was formless—naming the unnameable—we can begin to drain some of the venom that trauma injects into our veins. The act of writing, of telling, can allow us to read and reread our story until the words lose their sting, until they become part of the narrative of our resilience. And as we pen these experiences, we're more than observers. We're also creators. In the act of writing, we can assert control over a past that once held us hostage. We decide which details to include, which to omit, how to present our characters, and the light in which we want to paint our experiences. This, in itself, is a healing act.

Shortly after the Gatlinburg fire, Michael Reed began writing. It's a task as daunting as scaling a cliff face, yet the view from the summit—the clarity, the perspective—is often worth the ascent, he suggests. As we translate pain into words, we externalize it, witnessing it from a safer vantage point. We can study it as a naturalist would an intriguing specimen: with curiosity and compassion, but with a necessary detachment.

"There needs to be a national conversation about grief, loss, and mental illness," he says. "It is all still so taboo. There are millions of people out there who feel secluded, ashamed, embarrassed, abandoned, and unwanted. People need to know that it's okay not to be okay." Reed argues that writing about trauma is to understand that, as in nature, there is a season for everything. There is a season to hurt, a season to heal, a season to break down, and a season to build up. As we trace our narratives, the trauma that once seemed insurmountable becomes a chapter in our book of life—acknowledged and understood, but not the whole story. To write about trauma is to echo the rhythm of nature: to remember that after every winter, there is a spring; that after every night, there is a dawn. It is to recognize that we, like the Earth, are always in the process of becoming.

"This is the reason I'm still here," he says. "This is my purpose. I just don't know how to reach everyone who needs to hear it." He has a blog, but mostly he walks past people and wonders about their stories. "How amazing would this world be if we all wore our scars with pride instead of trying to cover them up? What if, by allowing our insecurities and vulnerabilities to be exposed, we could allow the world to see inside our souls to show who we truly are?" Stories are bridges, he argues. They span the abyss of loneliness that trauma often carves within us. By sharing our narratives, we can reach out to others. We say something like, *See, here is my journey. Here are my wounds.* And when those words find an echo in someone else's experience, they whisper of a shared human experience and weave a connection that reminds us we are not alone. Sometimes, above all else, this is what we need to hear and know to heal.

For Joyner, storytelling is both a passion and a profession. Her new endeavor is called Good Energy—a climate communications consultancy for Hollywood. She's trying to weave climate storytelling into blockbusters to ensure more people are aware of its reality. "There are reasons I don't love the hero's journey," says Joyner. "But one thing I do love about that kind of narrative arc is that the hero or heroine isn't doing it because they think they will win. Frodo didn't go up against Mordor because he thought he was going to win; he went up against Mordor because he thought it was the right thing to do."

She thinks storytelling is the right thing to do. Joyner has lived through three near-death experiences, and sometimes she offers up her own background as a means of climate communication. "So many people aren't in a position where they can share their vulnerable stories," she recently told her podcast cohost. "Because of my own privilege—to have a pretty secure job and a really supportive family—I am able to do that. I wanted to take advantage of that privilege to offer my story to the world in hopes that it does help someone." She sounds a bit like Reed.

Her first near-death experience came crossing the Tasman Sea from Australia to New Zealand. She was young—nineteen—and looking for adventure with a friend. They'd wanted to find something on the water, something they couldn't do anywhere else. They had a bit of sailing experience. They found a captain looking for a crew. They set sail. Eight days into the journey, though, a storm hit. It was five knots below hurricane force. "If you lose both of your sails in a storm like that, you turn over and you die," she recalls. They lost one, but they made it through the night.

Her second near-death experience came in September 2020, when Hurricane Sally hit her home in Alabama, where freshwater rivers and marshland meet salt water from the Gulf. "I did not take into account how truly stressful and traumatic living on the front lines of climate change actually is," she said. Sally had quickly intensified in the middle of the night, and what Joyner had gone to bed thinking was just another tropical storm woke her up at one a.m. as a Category 2 hurricane. The water was already within fifteen feet of her house. Her ninety-year-old grandfather lived two houses down and used a wheelchair. Over at his place, the water was already a foot below the windows.

Hurried packing. Phones, computers, favorite books, family jewelry. Little notes. They had less than two hours. She evacuated her grandfather in a foot of water. Driving out, it felt like being in a washing machine.

Later, surveying the wreckage, she called her home a war zone. "We lost twenty-four trees, including a huge, huge oak that was ancient and fell exactly where we evacuated my grandfather." She recalled it almost feeling like betrayal. She was already a climate activist.

Her third near-death experience came from the anxiety and numbing that followed. A friend had gotten her a hotel room, and in it, she started drinking, mostly to sleep.

But she couldn't. And then she really started drinking. "I basically drank 24/7," she recalled. "I was taking weed gummies. I was taking actual sleeping pills and anxiety medicine." She would sleep for an hour or two here or there. Ultimately, her doctor would suggest she was experiencing PTSD on top of a manic episode. (Joyner was already living with bipolar disorder.) "I was familiar with depression and manic episodes," she said, "but I had never had them together in a manner where I was deeply depressed and also manic. And it was utterly terrifying."

"I was just casually vividly imagining myself just walking out into the sea, and just disappearing."

Joyner's husband, who had been away, cleaning the wreckage, would immediately notice the difference in Anna Jane when he came to the hotel two or three days later. He got on the phone with her psychiatrist, and within a week, she was on a plane to a treatment facility. A staggering month would follow, but she came out the other side.

It's this message—of reemergence—she wants to share. "No one should be ashamed of the life that we're living in this world," she said.

There is a man sitting in a driveway. He is not sure why he has come here. It was December 2022, and Michael Reed had driven to his old home for the first time since it all burned up. It had been six years. He had inched up the road to the house and nosed his truck into the old driveway. And then he got out of the truck and sat down on the ground.

Before him was a different house, chalet-style and all wrong. He sat in the driveway, looking for a sign of Constance or Chloe or Lily. Sometimes they would appear to him as the animals they'd loved in life. But there were no ladybugs or owls or butterflies here. There was just a burned-out stump. He didn't stay for long.

Time to leave and focus on the case. "For me, this case is my only chance for closure," he tells me. "This lawsuit is directly related to my mental health. I could talk to you for hours about survivors' guilt. Anyone who says it isn't real has simply never lost someone close." As the case against the Park Service slowly worms its way through the court system, he wonders about accountability. Without it, he's left alone with his thoughts. "Our brains are more powerful than we realize. But when the brain begins to misfire and irrational thoughts become consuming, our brains can become our worst enemies," he says. He still has PTSD, but he's still writing. Some days are better than others.

"I don't know why God allowed 3/4 of my soul to be taken from me," Reed wrote in his blog around that time. "I don't know why I can't heal. I don't know why it hurts just as bad today as it did six years ago. I don't know why every holiday only gets harder. I don't know why I'm still afraid of fog." But I've never met anyone who's actually satisfied by the parable of Job. Michael, you've begun to tell us why you are afraid.

PART III

DISPLACEMENT

Sensing,
Pain,
Language

7

KARL FRISTON'S THEORY OF EVERYTHING

This is a much more serious and terrible experience, very different from the trip which you can enjoy if you know you took the LSD.
—Gregory Bateson, *Steps to an Ecology of Mind,* 1972

How can we know the dancer from the dance?
—William Butler Yeats, "Among School Children," 1928

magine you are a clown fish. A juvenile clown fish, specifically, in the year 2100. You live near a coral reef. You are orange and white, which doesn't really matter. What matters is that you have these little ear stones called otoliths in your inner ear, and when sound waves pass through the water and then through your body, these otoliths move and displace tiny hair cells, which trigger electrochemical signals in your auditory nerve. Nemo, you are hearing.

But you are not hearing well. In this version of century's end, humankind has managed to pump the climate brakes a smidge, but it has not reversed the trends that were apparent a hundred years earlier. In this 2100, atmospheric carbon dioxide levels have risen from four hundred parts per million at the turn of the millennium to six hundred parts per million—a middle-of-the-road forecast. For you and your otoliths, this increase in carbon dioxide is significant, because your ear stones are made of calcium carbonate, a carbon-based salt, and ocean acidification makes them grow larger. Your ear stones are big and clunky, and the clicks and chirps of resident crustaceans and all the larger reef fish have gone all screwy. Normally, you would avoid these noises, because they suggest predatory danger. Instead, you swim toward them, as a person wearing headphones might walk into an intersection, oblivious to the honking truck with the faulty brakes. Nobody will make a movie about your life, Nemo, because nobody will find you.

It's not a toy example. In 2011, an international team of researchers led by Hong Young Yan at the Academia Sinica, in Taiwan, simulated these kinds of future acidic conditions in seawater tanks. A previous study had found that ocean acidification could compromise young fishes' abilities to distinguish between odors of friends and foes, leaving them attracted to smells they'd usually avoid. At the highest levels of acidification, the fish failed to respond to olfactory signals at all. Hong and his colleagues suspected the same phenomenon might apply to fish ears. Rearing dozens of clown fish in tanks of varying carbon dioxide concentrations, the researchers tested their hypothesis by placing waterproof speakers in the water, playing recordings from predator-rich reefs, and assessing whether the fish avoided the source of the sounds. In all but the present-day control conditions, the fish failed to swim away. It was like they couldn't hear the danger.

In Hong's study, though, it's not exactly clear if the whole story is a story of otolith inflation. Other experiments had indeed found that high ocean acidity could spur growth in fish ear stones, but Hong and his colleagues hadn't actually noticed any in theirs. Besides, marine biologists who later mathematically modeled the effects of oversize otoliths concluded that bigger stones would likely *increase* the sensitivity of fish ears—which, who knows, "could prove to be beneficial or detrimental, depending on how a fish perceives this increased sensitivity." The ability to attune to distant sounds could be useful for navigation. On the other hand, maybe ear stones would just pick up more background noise from the sea, and the din of this marine cocktail party would drown out useful vibrations. The researchers didn't know.

The uncertainty with the otoliths led Hong and his colleagues to conclude that perhaps the carbon dioxide was doing something else—something more sinister in its subtlety. Perhaps, instead, the gas was directly interfering with the fishes' nervous systems: Perhaps the trouble with their hearing wasn't exclusively a problem of sensory organs, but rather a manifestation of something more fundamental. Perhaps the fish brains couldn't *process* the auditory signals they were receiving from their inner ears.

The following year, a colleague of Hong Yan's, one Philip Munday at James Cook University in Queensland, Australia, appeared to confirm this suspicion. His theory had the look of a hijacking.

A neuron is like a house: insulated, occasionally permeable, maybe a little leaky. Just as one might open a window during a stuffy party to let in a bit of cool air, brain cells take advantage of physical differences across their walls in order to keep the neural conversation flowing. In the case of nervous systems, the differentials don't come with respect to temperature, though; they're electrical. Within living bodies float various ions—potassium, sodium, chloride, and the like—and because they've gained or lost an electron here or there, they're all electrically charged. The relative balance of these atoms inside and outside a given neuron induces a voltage difference across the cell's membrane: Compared to the outside, the inside of most neurons is more negatively charged. But a brain cell's walls have windows too, and when you open them, ions can flow through, spurring electrical changes.

In practice, a neuron's windows are proteins spanning their membranes. Like a house's, they come in a cornucopia of shapes and sizes, and while you can't fit a couch through a porthole, a window is still a window when it comes to those physical differentials. If it's hot inside and cold outside, opening one will always cool you down.

Until it doesn't.

Here is the clown fish neural hijacking proposed by Philip Munday. What he and his colleagues hypothesized was that excess carbon dioxide in seawater leads to an irregular accumulation of bicarbonate molecules inside fish neurons. The problem for neuronal signaling is that this bicarbonate also carries an electrical charge, and too much of it inside the cells ultimately causes a reversal of the normal electrical conditions. At the neural house party, now it's colder inside than out. When you open the windows—the ion channels—atoms flow in the opposite direction.

Munday's theory applied to a particular type of ion channel: one responsible for *inhibiting* neural activity. One of the things all nervous systems do is balance excitation and inhibition. Too much of the former and you get something like a seizure; too much of the latter and you get something like a coma—it's in the balance we find the richness of experience. But with a reversal of electrical conditions, Munday's inhibitory channels become excitatory. And then? All bets are off. For a brain, it would be like pressing a bunch of random buttons in a cockpit and hoping the plane stays in the air. In clown fish, if Munday is right, the acidic seawater appears to short-circuit the fishes' senses of smell and hearing, and they swim toward peril. It is difficult to ignore the question of what the rest of us might be swimming toward.

Sensing the world is the most important part of making sense of it. No organism could make judgments and decisions relevant to its survival if it didn't have a mode of collecting information about what's out there, beyond itself. Sensory systems—like sight, touch, and smell—offer a means for animals to sample their environments and detect changes in them. Without noticing these changes, without sensing them, no living thing could adjust its behavior in a manner that ensures its responses to danger (or pleasure) are appropriate as opposed to random. Without access to our senses, we'd just be hockey pucks, careening about and waiting to be slapped.

Sensory information, the stuff we sense, is composed of multiple impinging signals. Photons bounce off the dense mosaic of atoms we call the environment; different wavelengths of light carry with them different levels of energy. Creatures and smokestacks and landfills and apple slices emit gaseous molecules of myriad composition. These same environmental inhabitants—the living and nonliving around us—often vibrate in such a way that they compress and rarify little pockets of air; at sea level, these invisible waves travel at 340 meters per second. Things can smell rotten or be magnetic. Things can have textures.

In the interest of interpreting the world, it is useful for organisms to have a physical place to represent all this information. In people, much of that work happens in the brain. You possess fairly specialized areas for representing the data collected by your sensory organs. The occipital lobe, for example, a small chunk of brain matter in the back of your head, is responsible for processing visual information. The auditory cortex is tucked into cerebral fissures on the left and right sides of your brain.

In the 1950s, when neuroscience was beginning to come into its own as a field, early electrophysiological experiments often amounted to sticking electrodes into the visual cortex of cats and watching what happened when various patterns of light were shone on the animals' eyes. It is from these experiments we get anecdotes like the following:

> *Suddenly, just as we inserted one of our glass slides into the ophthalmoscope, the cell seemed to come to life and began to fire impulses like a machine gun. It took a while to discover that the firing had nothing to do with the small opaque spot—the cell was responding to the fine moving shadow cast by the edge of the glass slide.*

One example among many, from the Harvard neurophysiologists David Hubel and Torsten Wiesel, who via this case gained an understanding of orientation- and direction-selective cells in the primary visual cortex: individual neurons that responded not just to light, but to its physical qualities as it moved through space and cast shadows on objects. Mid-twentieth-century neuroscience often looked like Hubel and Wiesel's delicate poking around. (The pair would win a Nobel Prize in 1981, in part for their serendipitous discoveries about cat brains.)

But most of the brain doesn't do this kind of thing. The primary sensory cortices are reasonably interesting and easy to study; much of the cerebral surface of the brain, in contrast, is concerned with *combining and comparing* primary sensory information and other neural data. And these cortices are much more mysterious. They are complicated enough to be shrouded in amorphous euphemism: the "association cortices." Contemporary neuroscience textbooks tend to stumble through a definition that looks, to pull just one example off the shelf, something like this: "The diverse functions of the association cortices are loosely referred to as 'cognition,' which literally means the process by which we come to know the world ('cognition' is perhaps not the best word to indicate this wide range of neural functions, but it has already become part of the working vocabulary of neurologists and neuroscientists)." It is not the kind of statement that inspires confidence in our collective knowledge of the areas under investigation. But we do know some things about the association cortices. We know, for example, where they receive neural information from—areas like the primary visual and auditory cortex, but also the motor cortex, the area responsible for coordinating our movements—and we know where they send it once they've evaluated the signals they've received. Once neural data are processed in the association cortices (associated, as it were), these brain regions broadcast signals to internal relay centers and motor pattern generators like the thalamus and the basal ganglia, as well as to memory areas like the hippocampus. In broad strokes, the association cortex often appears to be collating a multisensory picture of the world and then figuring out what to do with it. It is turning the sensing into sense making.

The rub here is that the multisensory picture in question is not static. Brains exist in time, and the world around them frequently changes in one way or another. In order to know what is going to happen next—at the next step in time—we need to constantly sample and reassociate the world. We need to have some idea of what happened in the previous moment and some idea about what we think might happen in the next. Without these temporal connections, we would live in a constant state of bewilderment. Every instant would be a surprise. And not in the birthday-party sense. These surprises would be existential in nature: Every moment, you would need to relearn that gravity exists; that your arms are connected to your body; that the world isn't actually disappearing when you blink. It would be worse than not having a memory. You wouldn't even know what to do with a memory if you had one.

In other words, for you to sustain your existence, your internal model of the world must minimize surprise. We're not operating in the realm of the metaphorical here. From moment to moment, nervous systems sample the environment using their vast array of sensors, including tools like eyes and ears and noses. In places like the association cortices, brains then compare this stream of perceptual information to what they'd predicted they were going to receive, given their current understanding of the state of the world. They compare these expectations to the new percepts, and then they adjust the world model as a function of the difference—the model's error—all with the aim of minimizing surprises that might arise during the next comparison.

Seen in this way, the job of any organism is to collect evidence for its own model of the world. That's how it survives. And if the goal of the agent's modeling efforts is to best represent the outside world—to best generate hypotheses that explain surprising events—then the version of the model that is able to do so is *also* the version of the model with the best underlying evidence. So here is the cymbal crash we need to get to after all that drumroll: Minimizing model prediction error is the same thing as maximizing model self-evidence. And if organisms *are* indeed models of their environments—if what they do is represent the world in order to navigate it—then that's the same thing as saying that "minimizing surprise is the same thing as maximizing the sensory evidence for an agent's existence."

It's okay to read that last quote a couple dozen times—it's a bit of a brain bender. It comes from a University College London neuroscientist named Karl Friston, who has spent much of the past two decades developing, expanding, and championing a unified brain theory rooted in surprise minimization. It's a little difficult to write about surprise minimization and existence and not sound a little New Agey. But the difference between Karl Friston and Deepak Chopra is that Friston has the math to back up his theory—in fact, it's all math—and, more importantly, the implications of his predictions have been borne out empirically. Friston was considered for a Nobel a few years ago, and for some, his work on surprise minimization is likely the most profound effort that academic neuroscience has seen in half a century. But, yes, if the prior two paragraphs left you a pinch mystified, you are not alone. In 2022, *The Lancet* called Friston's work "famously inscrutable."

We need to scrutinize it, though, if we want to understand the stakes of acidifying oceans and Philip Munday's clown fish. Consider what we've addressed thus far. Namely, it would be irresponsible of the brain to pretend the world isn't changing. We've already seen as much, in the first chapter, on memory. If we didn't forget the world of old, we wouldn't be able to adapt to the dynamic present. Friston's theory formalizes this logic at multiple scales of analysis. It says the point of all living things is to maximize evidence for their own model—their own existence—to minimize the surprise they encounter in the world, relative to their current understanding of the relationship between causes and consequences. The causes, here, amount to reality:

the everything outside ourselves. The consequences are our perceptions of this world—the sensations we feel when navigating our environments. When our expectations match our perceptions, our brains are doing a good job. When they don't, we adjust accordingly. We adapt.

But what if environmental changes beget more than just model adjustment? What if, instead, they're corroding the ability of our models to adapt at all? Our modeling efforts depend on our ability to sense the environment. But if climate change is coming for our very senses—as researchers' work on ocean acidification and rainfall patterns and warming has shown—the news is doubly bad. It's not just that animals' brains change with a changing climate; it's that a changing climate also threatens our ability to notice some of the signals most relevant to our survival.

The effects are stealthy and subtle. Forget the ear stones and the ion channels for a moment. Ocean acidification in its own right also increases the degree of ambient noise in water, since sound absorption is a function of water pH. Warming and acidification can push marine animals deeper into the ocean, where it's harder for them to see. On the shore, researchers have identified the disruptive effects of carbon dioxide on olfactory-system cascades in crustaceans. Farther up the hill, altered temperature and rainfall patterns can affect the ability of insects to detect and respond to chemical cues. Outside the climate arena, noise pollution bears on marine animals' ability to communicate and navigate. Pesticides can impair the homing abilities of honeybees.

Karl Friston is silver-haired, with a furrowed brow balanced by an easy smile. When I reached him via video call in his home in London, he told me that the impacts of environmental change on the sensing brain lend themselves, effectively, to holding false beliefs with higher certainty. "This impact could be read as a delusion," he said. "In effect, the thing you're trying to accumulate evidence for is itself compromising the ability to update the model." It wasn't an alien idea for Friston, despite his not frequently working on climate change. Compromised models arise all the time in neuropsychiatric research. When he'd previously studied schizophrenia with colleagues, the notion of "decreasing confidence in sensory input" was that which ultimately offered one of the best explanations for how delusions might arise in the first place. "The rate at which I change my beliefs depends very sensitively on estimating the precision of the evidence at hand correctly," he said. And if that precision degrades, organisms become rightfully and rationally resistant to change. I mentioned the clown fish, losing their senses at high temperatures and at high levels of ocean acidification. "Effectively," he said, "the fish is becoming delusional."

When you see a clip of a vase breaking in reverse—of shards reassembling—you know something is off. That's not how the world is supposed to work, right? Indeed, it's not how the world works at all. The flow of time dictates it can't. When you see the vase reassemble, your knee-jerk disbelief stems from the fact that the arrow of time is pointing in the wrong direction. Thermodynamically speaking, the wrongness corresponds to a violation of a fundamental law of nature: Namely, that in the long run, the universe tends toward randomness. Glass shatters. It does not unshatter. This universal attraction to chaos—to a measure of disorder, called entropy, to be precise—is, in fact, the second law of thermodynamics. It is what points the arrow of time. When thermodynamic systems (like the vase) appear to us as becoming more disorderly, we understand the experience as time's arrow pointing toward the future, as it normally does. If we observe the vase reassembling, if entropy is decreasing, the arrow must be pointing toward the past. "That is the only distinction known to physics," wrote the British astrophysicist Arthur Eddington in 1927, when he developed the notion of time's arrow. "This follows at once if our fundamental contention is admitted that the introduction of randomness is the only thing which cannot be undone."

Christopher Nolan has not admitted Eddington's contention. In his 2020 film *Tenet*, the arrow of time is something of a Pollyannaish suggestion. Early in the plot, viewers are introduced to the idea of an entropy-inverted bullet: a ballistic that travels backward, temporally speaking. As Clémence Poésy's character offers to a confused protagonist (played by John David Washington): "You are not shooting the bullet, you are catching it." The perspective shift is essential for identifying the film's baddies, who are inverting the entropy of all kinds of things. Washington's character must pay attention to the arrow of time if he is to get a grip on reality; on causality; on the order of things. *Inception*-like antics ensue.

Enter stage left Morten Kringelbach and Gustavo Deco, neuroscientists at Oxford and Spain's Pompeu Fabra University, respectively. (We're in reality now, not the film.) Kringelbach and Deco are interested in the nature of life—and in particular, in the statistical duct tape that keeps it from unraveling. Like Karl Friston, they seek to understand the fundamental principles that allow thermodynamic systems to sustain a kind of nonequilibrium, since, as they write, "the ultimate equilibrium is death." How is it that the environment and the brain interact to avoid this fatal equilibrium? To answer as much, the neuroscientists have done something almost Nolan-like in their labs. They have caught the arrow of time.

Recall that systems with increasing entropy respect the arrow. You can't put the glass back together. But not all thermodynamic systems are like this. "In contrast," write the neuroscientists in a recent essay, "as an example of a system in equilibrium, imagine watching footage of colliding billiard balls. When watching this film both forwards and backwards, you would be hard pressed to distinguish the arrow of time for each film. In thermodynamical terms, this is because the process is not producing entropy and creates an intrinsically reversible process." Entropy-producing systems are nonreversible. Zero-entropy systems are the opposite. Okay, good. But pay attention to the inverted frame here, how the researchers are relating time and entropy. As per Eddington's original formulation, you can understand the arrow of time as describing the entropy production of a system. But you can also flip the equation around and examine the entropy production of a system *in order to understand the arrow*. There is a symmetry around the equal-sign: Time helps us get a grip on disorder, but disorder also helps us get a grip on time. And because both of these elements are intimately tied to the question of sustaining nonequilibrium—of staying alive—it occurred to Kringelbach and Deco to poke around for them in the cranium. Understanding the reversibility of the brain could help them understand how the environment and the body might be influencing its mechanics. They call their new field the thermodynamics of mind.

Kringelbach and Deco's system for teasing apart these factors is an algorithmic approach to brain analysis called the Temporal Evolution NET (TENET). The neuroscientists feed TENET a variety of electrical signals recorded in the brain—and, crucially, versions of such signals that have been artificially reversed in time—and teach the algorithm to differentiate between forward signals and backward signals. When the training is complete, the researchers show

TENET new examples of brain electrical activity and ask it to characterize the mind's reversibility—and by extension, its arrow of time. The feat is possible because different brain states are associated with different levels of disorder. By understanding this entropy production and its relationship with the broader environment and the rest of the body—by understanding what's causing what—Kringelbach and Deco can estimate with unparalleled precision the extent to which the environment may be driving the brain toward nonequilibrium. And, yes, the algorithm's name is a reference to the movie.

In 2022, the researchers confirmed that "in general, the brain is being driven by the environment and, importantly, the human brain is closer to non-equilibrium and more irreversible when performing different tasks than when resting." In other words, in no small manner, perceiving and interacting with our environment keeps us alive, because the relationship prevents us from collapsing into a steady state of equilibrium. Strikingly, though, when examining brain recordings from patients with ADHD, bipolar disorders, and schizophrenia, the researchers noted that the brain's resting states were closer to equilibrium than those of people without these conditions. The finding illustrates that "the brains of neuropsychiatric patients are more isolated from the environment and more likely to be intrinsically driven," they wrote. "This fits with how, for example, rumination in depressed patients can lead to the malignant isolation from the external world that can drive depression."

Their speculation is in line with Karl Friston's musings on neuropsychiatric disorders and surprise-minimizing models. By definition, "if the world changes, the generative model has to change," Friston told me. That's the whole point. Importantly, though, that change *doesn't* always occur—or, indeed, when it does, it isn't always healthy. Sometimes the manner of coping with a change can be pathological.

Consider the case of depression, as Kringelbach and Deco have done. In Friston's framing, depressive isolation is a perfectly rational manner of coping with a frightening, crushing, changing world. The problem is that this isolation is also self-reinforcing: It's hard to get out of the rut. "Certainly, one way of coping with that world is to withdraw from it," said Friston. "It's perfectly functional avoidance behavior. But you sacrifice the opportunity to test the hypothesis that the world is safe." In the thermodynamic sense, it's as if the mind is trending closer to an equilibrium state. "You've become locked into a self-affirming generative model. It maintains itself because you don't look anywhere else."

For Friston, neuropsychiatric conditions like depression and schizophrenia have much to teach us about the brain. But we can read in them deeper understandings of our relationship with the world too. From the perspective of surprise minimization, for example, the opposite of the aberrantly self-reinforcing model is that which interacts dynamically and specifically with the environment in order to cope with change.

At first blush, this last idea is almost so simple that it feels like a truism. But its beauty rests on a critical understanding of what this whole "environment" business is, anyway: this big reality outside of the model. Surely, *the world* is its own best model. What game are we playing attempting to represent it in the first place?

There are two shining ecological diamonds buried in Friston's theories of organism–environment interactions. From the surprise-minimization perspective, the world-as-model idea is wrong. Instead, *the world is its own best world*, and it's the organism that constitutes the model. Crucially, it is a selective model. It can't be a full model of the world; otherwise it would *be* the world. Instead, the organism is "the best model of those aspects of the world relevant to its surviving and thriving—a familiar econiche that it has largely constructed for itself." Seen in this way, evolution is a mode of surprise minimization too. That is a shining ecological diamond for us to ponder.

And if you're comfortable broadening the lens that far, past the organism and to the species, you should be ready for the second one. These econiches, the carve-outs of the environment to which we're well suited—they don't just work for us; they work with us. Consider elephant paths, the practiced routes through grass that animals (including people) tend to take. We trace paths through fields as a matter of habit, of following the footsteps in front of us. But there's a Fristonian reading here. We can also understand elephant paths *as the field perceiving a pattern*—as modeling the creatures walking the paths. Footsteps form depressions (a multivalent word). So here: The glimmer at the bottom of the mine, the true deep beauty nestled into Karl Friston's theory, is that surprise minimization is perfectly reciprocal. Mathematically, it doesn't matter which side of the organism–environment divide you're on: There are sensations passing in one direction and actions passing in the other. We tend to understand ourselves as perceiving the world and then acting on it in kind. The semantic flipside is just as true: Our actions are the world's sensations, and our sensations are the world's actions. It's all a dance.

don't offer these ideas to encourage us to get lost in the dance hall, though. I offer them as a means of inviting us to pay closer attention to the shifting sensuous landscape around us—and to what its perturbations might mean for animals (including animals like people), as well as the intrinsic value of these shifts in their own right.

In 2019, a curious paper tumbled out of the academy; "Climate Change Is Breaking Earth's Beat," proclaimed the title. Led by Jérôme Sueur of the Institut de Systématique, Évolution, Biodiversité at the Sorbonne's Muséum National d'Histoire Naturelle, the team of researchers argued that an underappreciated effect of climate change stemmed from its ability to alter the "natural acoustic fabric" of Earth. "Here," they wrote, "we emphasize that climate change is transforming two components of the global soundscapes: biophony which includes biotic sounds; and geophony that groups abiotic, but natural, sounds coming, for example, from running water, wind, and earth movements."

They pointed to the fact that sound speed depends on environmental factors like temperature, humidity, wind, and rain intensity; that climatic shifts can spur changes in the seasonality of a landscape's acoustic features (like birdsong); that as whole species are lost and gained, we can come to expect "an acoustic homogenization, with ubiquitous species dominating the soundscapes." Just as Friston offers the notion of an econiche to understand what, exactly, it is that species are modeling, Sueur and his colleagues point to the notion of an acoustic niche to suggest what we stand to lose in the face of climate change. "The acoustic niche hypothesis," they write, "derived from the ecological niche concept, stipulates that at a given location each species would occupy non-overlapping specific acoustic space to avoid interference." With climate disruption—and its subsequent effects on biodiversity and physical space—these partitions are threatened. And who can say what the ensuing ripples of asynchrony might perturb? These are uncharted waters.

Already we're hearing some of these waves, though. In Ithaca, New York, back in 2001, environmental scientists James Gibbs and

Alvin Breisch delved into the New York State Amphibian and Reptile Atlas Project, a dense biological compendium that included, among many other juicy tidbits, records on seasonal timing of frog calls over time. They compared these data to archival records of the same from the early twentieth century, under the hypothesis that, because temperature can influence reproduction in frogs, the timing of their calls might offer "a sensitive index of biotic response to climate change." And it did. In the Ithaca area, at least four frog species were calling a week and a half to two weeks earlier in 2000 than they had a hundred years previously. As the world warmed, the aural landscape was changing—and the reproductive landscape along with it.

Color, too, is subject to a changing climate. As temperatures soar, plant life the world over faces an uphill battle, as greenery grapples with heat waves and the prolonged droughts that are becoming increasingly common. Vast swathes of foliage can shed their vibrant hues in exchange for parched yellows and subdued browns. Evaporation and dry soil can leave plants vulnerable to extreme desiccation. Accordingly, leaf stomata, plants' tiny pores responsible for regulating gas exchange, can constrict to minimize water loss. This self-protective mechanism compromises the efficiency of photosynthesis. In particular, plants tend to respond by reducing their production of chlorophyll. Chlorophyll is what makes plants green; somber hues emerge in turn. Wildfires leave behind charred landscapes. Melting snow and ice leave behind darker terrain and exposed ocean.

The changing of the world's sound profile; the molting of some of its color—maybe these perceptual vectors of global warming aren't as dire as cyanobacteria-fueled neurodegeneration or post-traumatic stress disorder. To lose a little green isn't the same as losing a loved one. Or is it? Surely it is okay to love a color and to grieve its dimming.

For Friston, it all boils down to sensation. And in this manner, grief is always a mode of learning about the world. Loss is evidence that the world has changed. When a color dims, we have to adapt to this new environment—just as we must when someone close to us is taken from our everyday lives. It might take us years to understand what has happened. "The common thread here is the manner in which we adapt to surprising events in a way that's self-maintaining," he said. Your model "has to explain the sensory evidence over the coming weeks and decades. It's true whether you're losing your seasonal fluctuations or a loved one."

In 1891, in Baltimore, the Kennard Novelty Company patented a new toy. It ought to have represented a welcome development for children—at least the affluent ones—for whom the nineteenth century was a time of dolls and marbles and wooden blocks. But Kennard's new toy was geared ever so slightly more toward adults—toward the spread of spiritualism sweeping the country and toward the reckoning with postbellum losses still felt by grieving families. All told, it wasn't much to look at. Heck, you could have mistaken it for a cutting board. But it was special, nonetheless, because when you played with it, you could communicate with the dead.

Ouija boards worked—and still work today—because of a phenomenon known as the ideomotor effect. Introduced by Victorian psychologists in the 1840s, the effect refers to any process in which thoughts or mental imagery appear to spur an unconscious or reflexive movement. When analytical hypnotherapists interpret physical movements as answers to questions, they are leveraging ideomotor effects in their patients. Some lie-detection software depends on such effects: Minuscule, unconscious hand movements can make up what's known as a "tell." All-star poker players claim to be able to recognize one another's tells. (In the 1998 poker film *Rounders*, the mobster Teddy KGB, played by John Malkovich, eats an Oreo every time he has a good hand.)

The ideomotor effect comes in various flavors of pseudoscience. But it also holds a kernel of truth. So-called ideo-dynamic responses are everywhere in the body: Any instance in which a thought spurs a bodily reaction falls under this broader umbrella. If you salivate when you think of your favorite food, you are experiencing an ideo-dynamic response. You may be doing as much right now. You don't *choose* to exhibit reflexes. Generally speaking, you don't *choose* to secrete anything. But your thoughts can spur those types of reactions nonetheless.

There is a grand secret here, another diamond, maybe, tucked away in the folds of this disconnect. In particular, via the framework of surprise minimization, we can consider *all* of our actions to operate in this manner. The principle depends on a kind of sparsity.

Recall that we are in the game of minimizing prediction error. Our sensory systems collect information about states of the world, and then we compare this information to what we'd expected to have happened in that moment—we think; we interpret. Then . . . what? We act. And finally, we observe the consequences of our actions. We sense them, and then the cycle repeats.

Pay close attention to what just happened. In this chain of events, our actions follow directly from a calculation of a prediction error. Actions are just as much a part of our model as everything else in our noggins, which is also to suggest that our actions are similarly in the business of minimizing surprise. Importantly, though, *it wasn't our senses* that caused our actions. It was the prediction error. There's an air gap.

What this means is that for you to live sentiently—for a perception–

action cycle to *work*—the math of Karl Friston tells us that there's no direct connection between your actions and your perceptions: You don't, for example, feel yourself directing your hand to pick up a glass of water. You might experience the *intention* to do so, but then you kind of just ghost toward the glass, and you perceive yourself grasping it with your eyes and your fingers and your proprioceptive system. You only ever perceive the *consequences* of your actions, and you have to infer the causes of those consequences, just like you have to infer the causes of all the other worldly consequences forming sensory impressions in your brain. You compare your predictions to your impressions, and you begin the cycle again. Your actions aren't directly wired to your senses. That's the sparsity. You are always learning your body's place in space, and your actions work to minimize your surprise about what you learn.

This is the ideomotor effect. Our awareness of our actions, in a very literal sense, is only ever a function of us *observing* those actions. Our perception works to optimize our predictions in a manner that allows our actions to minimize surprise. Accordingly, our actions are only ever a function of our beliefs about what will happen next. And when these phenomena work together, in synergy, we avoid potential danger. We stay within our econiches.

It is a beautiful implication of Karl Friston's work: that the conditions that make life possible are defined by a kind of sparsity. As he wrote in 2017, surprise minimization effectively formalizes ideomotor theory:

> *And suggests that all movements are prescribed by mental images that correspond to prior beliefs about what will happen next. These priors are inherently dynamic and itinerant. This suggests that our exchanges with our environment are constrained to an exquisite degree by local*

and global brain dynamics; and that these dynamics have been carefully crafted by evolution, neurodevelopment, and experience to optimize behavior.

Poker tells aren't a footnote to human behavior; they are the key to the whole shebang.

But sparsity isn't the entire story. The conditions that make life possible are also defined by a kind of jitter. This jitter is related to the constant state of "dynamic and itinerant" learning in which we've found ourselves. You're always updating your model of the world, right? You're always exploring. You move, in some fashion or another: You wander down a forest path or you crane your neck to get a better glance at your neighbor or you flit your eyes around in little micromovements to keep the interesting parts of the world splashing onto your fovea. You move around so you don't get stuck in a rut. You jitter.

Importantly, though, these modes of exploration aren't infinite. Your jitter is bounded. You can't, for example, remove one of your eyeballs and squish it into a disklike shape and slide it under a door to see what's on the other side. You can't teleport, and you don't spontaneously combust. You have limits. For your model to learn the most useful things about the world—for your model to deliver on its promise of self-evidencing—it needs a *bounded* set of states it can inhabit. It can't be everything everywhere all at once. It also can't just be one thing all the time, or you wouldn't be very interesting. In fact, you would be dead. Another good word for "jitter" might be "movement."

Taken together, the principles of jitter and sparsity offer an unorthodox conclusion: Sustainability—literally, the condition of sustaining existence—depends on uncertainty and limited connection. And thus, we have arrived at the point of all this theorizing. One doesn't often hear activistic cries for limiting our relationships with one another. But Friston is serious and literal in his interpretation here. "If we're not careful about maintaining disconnection and sparsity, there are real dangers," he told me. "Connectivity is the killer."

Complex systems collapse when they sink into a fixed point or when they begin to oscillate uncontrollably and unravel. Think of climatic tipping points; of not being able to come back from the edge. Think of the Tacoma Narrows: the so-called Galloping Gertie suspension bridge that, in 1940, tumbled into Puget Sound when high winds spurred self-reinforcing oscillations in the deck. There is hardly a better example of something "spiraling out of control." Friston thinks that globalization is like this, that the website formerly known as Twitter is like this. "Globalization and boy-racer notions are dangerous and misguided," he argued—not necessarily because they tend to be extractive, but because *they often amount to one person influencing too many other people.* Outsize influence is the opposite of jitter. It is the opposite of randomness. Outsize influence is resonance and oscillation. It is the thing that causes complex dynamical systems to falter. "Anyone promoting a deglobalization agenda; they make me feel warm inside," said Friston.

We are getting a little speculative and abstract here, but I'm not sure what else we're supposed to be doing when we poke and prod at the meaning of life. And in some sense, that *is* what we're doing. Sustainable self-evidencing—that's it. That's the meaning, if there's one to be dug up. That is what it means *to be.*

So bear with the speculation and abstraction. The magic of surprise minimization is that it offers the tools for reading the dynamics of nature in the same language at every scale of analysis. You get to use the same words to talk about the behavior of a mitochondrion and a cell and an organism and a species and an ecosystem. If that sounds a little squishy, it's because we're not used to thinking this way. It beggars belief that the same mechanics of sustainable existence apply to the lonely brain cell and to the individual bonobo and to the whole population of chanterelle mushrooms in the forest. Friston's work is not about blithely dissolving the boundaries between individuals and arguing for some kind of "we are all connected" hand-waving. Quite the opposite. Surprise minimization *defines* individual existence, after all. It needs individuality to make any sense. Yes, sure, we are all connected. But that's boring. What's not boring is the fact that despite the porous boundaries between us, and in addition to the interconnectedness of the individual and its environment, *we also share a definition of success with every other thing that physically exists, at any scale.* "Sustainability" means the same thing for everything.

So with that in mind, let us ground ourselves. Apply Friston's values to sustainability as commonly understood in the environmental movement. How do we sustain the complex system that constitutes a livable planet? I'd offer that part of the goal here ought to be: become a little unpredictable. We have to be jitterbugs. That's what actual land reciprocity looks like right now, doesn't it? Movement. That's real climate empathy. You want to mirror the environment? Be a hurricane, and be a hurricane where they least expect it. Blow up a pipeline. Throw the dang soup.

The philosopher Timothy Morton (known colloquially as "the dark prince of the Anthropocene") recently offered a full-throated defense of tossing soup on a Van Gogh. "People become habituated to the soup, as it were, that they live in," they told *Time*. "That's the point of [the soup protest], to make everything suddenly uncanny . . . deliberately or not, to stop people and make them see things differently." Morton had an evidentiary dog in the fight: Soup had moved the needle for them. They'd spent six years trying to ban gasoline-powered leaf blowers in their neighborhood, and five hours after they'd tweeted a soup protest to the relevant body—which also controlled a nearby art collection—the leaf blowers were out. (Morton, who teaches English at Rice University, has since joined Just Stop Oil—the soup throwers in question—as a kind of art-directing adviser.)

"I didn't understand it at first," they said in a recent interview with the philosopher Chris Julien. "And then I realized: 'Oh, this feeling—that I can't understand it—is the point.' Activism at that scale is trying to interrupt you, to stop you, with something incomprehensible. The most beautiful part of that action is the meaninglessness."

To be clear, I don't think you literally need to throw soup at a Van Gogh to be the change you want to see in the world. In fact, I recommend against it. Not because the tactic is ineffective at getting gasoline leaf blowers banned—quite the opposite, by all measures—but because it has already been done. Today, soup tossing has reached the realm of predictability. You, on the other hand, need to be unpredictable. "Soup on a Van Gogh painting is multivalent. It has a ball bearing in the middle; and what is this ball bearing? It's total nonsense," said Morton. "Whoever controls the nonsense controls the world." You need to be weirder and goofier than the soup. With any luck, you'll surprise yourself.

8

BURN SCAR

If I can learn to love death then I can begin to find refuge in change.
—Terry Tempest Williams, *Refuge*, 1991

What would it look like if we allowed more and more things to have some kind of power over us?
—Timothy Morton, *Being Ecological*, 2018

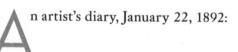An artist's diary, January 22, 1892:

One evening I was walking along a path, the city was on one side and the fjord below. I felt tired and ill. I stopped and looked out over the fjord—the sun was setting, and the clouds turning blood red. I sensed a scream passing through nature; it seemed to me that I heard the scream.

A stark entry, perhaps, given the glory of the painting that would immortalize the blood clouds. You are familiar with the work: It is called *The Scream*. The image of a wispy figure clutching their head, mouth slack, the sky on fire beyond. The pose is ubiquitous enough, the title synonymous enough with the act itself to have inspired an emoji. But read Edvard Munch's diary entry, and notice that it is not the figure in the painting that is doing the screaming. All we have there is an open mouth. The original German title for the piece is clarifying: *Der Schrei der Natur* (*The Shriek of Nature*). A screaming comes across the sky. The subject covers their ears.

Munch was Norwegian and living in Oslo (then Kristiania) at the time he encountered the crimson clouds that so disturbed him. To-day, his painting is the modern likeness of anxiety. But before he ever put tempera to cardboard, his experience on the bridge was just that: a solitary, personal experience. It was profound and debilitating. Munch would write poetry about the moment. He would paint four versions of *The Scream*. In its sister painting, *Fortvilelse* (*Despair*), a melancholic figure embodies the grief he'd cited in his journal—the figure leans over the railing, as a weary Munch had done that day. *Despair* takes as vantage the same point on the same bridge, as does a second sister work, the crowd-filled *Angst* (*Anxiety*). A friend of Munch's recalled the painter's fixation on the experience: "For a long time he had wanted to paint the memory of a sunset. Red as blood. No, it was coagulated blood." Something had happened on the bridge. A spasm of nature, tempestuous enough to drive Munch to obsession.

Many have scoured astronomical and meteorological records in search of the coagulated blood. Perhaps Munch had come across an instance of the striking polar stratospheric phenomenon known as nacreous clouds, which form from ice crystals and bear an undulat-ing resemblance to the painting's sky. Maybe it was just a normal sunset after all, but one punctuated by the sounds of the nearby slaughterhouse—or by the knowledge that Munch's sister was a resident at the asylum at the base of the hill. Maybe. But some scholars think the blood-sky phenomenon corresponded to a volcanic sunset: that the moment recalled in his journals and reified in so many poems and sketches had occurred a few years earlier, in the winter of 1883–84, when the gases and dust of the great Krakatoa eruption spurred fiery sundowns the world over. Despite the volcano being half a globe away, astronomers' contemporaneous records from the Kristiania Observatory note a "very intense red glow that amazed the observers" and that morphed into a "red band." In Munch's first study for *The Scream*, the only color he applies is the blood red to the knifelike clouds; the rest is just a charcoal sketch.

And maybe the precision of the genesis story doesn't matter too much. But what we do know is that the painting does not show some unbreachable, walled-off person, suffering internally without context. There is something happening outside the figure's head. In this manner, *The Scream* really does very little to prefigure, say, the contemporary stock-photo result for "anxiety"—some lone, seated woman, hands over her eyes, wallowing against a nondescript wall. In *The Scream*, the figure is porous and wavy. They lack definition. They mirror the clouds. Munch's point was always the screaming in the sky—and what it is capable of shattering within us.

It is no surprise that when the world breaks, we break along with it. The environment—the ground, the horizon, the sky, as perceived by all our senses—is our reference point. In some sense, as Karl Friston's work has shown, to exist at all is to model the world. The point of *being* is to keep this modeling effort chugging along. You are a mirror. You are a thousand mirrors, each one flitting to and fro to capture the scattered light of the world and focus it to a point. Sometimes this light shines from the faces of your friends or from the glow of a feast or the warmth of a hearth. But sometimes it comes from a Krakatoa, or from a killing blaze in Gatlinburg, or perhaps from an open pit where a mountain used to be, or a pristine childhood lake now host to a cyanobacteria bloom. It does not make sense for an embodied mind to assign the latter events a feeling of gladness. The mind has other tools to represent catastrophe. For this reason, angst, despair, and anxiety have long mirrored a fractious Earth—long before anyone coined the term "climate anxiety."

You have heard this latter phrase. It has come to mean something like *an overriding, near-pathological state of worry about the impending climate crisis*, and it has come to be applied with everything from eye-rolling disdain to self-righteous proclamation to genuine psychological and psychiatric concern.

What is climate anxiety? It is easier to suggest what it isn't. Climate anxiety is not what you feel when you experience a hurricane. It is not post-traumatic stress disorder. It is not generalized anxiety disorder, the term for the diagnosable condition we colloquially refer to as "anxiety." Climate anxiety is not yet a diagnosis at all, in fact; it does not appear in the *DSM-5*. Which is not to suggest the condition isn't real. Of course climate anxiety is real. But the phrase is a misnomer. Anxiety, psychiatrically, suggests irrational fear. There is nothing irrational about worrying about collapse.

In a recent psychological study of British Columbians who lived through the June 2021 heat dome, Canadian researchers concerned with this sort of thing had the opportunity to test out a new survey instrument. The Climate Change Anxiety Scale (CCAS), designed by psychologists in Ohio as "a measure of climate change anxiety that would allow for consistency in measurement and understandings," acts as a mental thermometer for negative feelings associated with climate change. Most of the study's participants in British Columbia "indicated that they were much (40.1%) or somewhat (18.4%) more worried about climate change due to the heat dome." On average, post-scorch, their CCAS scores jumped from 1.66 to 1.87: an increase worthy of publication in *The Journal of Climate Change and Health*, and a discrete, if lukewarm, example of climate concern ticking up in the wake of extreme weather.

When the Ohio psychologists designed the CCAS, they'd sought a tool that would allow them to compare human reactions to climate change over time. They'd also wanted lexical clarification. "Most fundamentally," they wrote, "a valid measure also allows us to define what it is we are talking about when we talk about climate anxiety."

We discussed what the British Columbians may have been talking about in the second chapter, on cognition. During the heat wave, the mercury in a town called Lytton had reached 121.3 degrees Fahrenheit (49.6°C). Nowhere in Europe—or, for that matter, South America—have meteorologists ever recorded a temperature that high. A day after setting the record, Lytton burned to the ground. It flamed out. On the Pacific Northwest coast, crustaceans baked in their shells and the beaches smelled like gut rot. When I visited one of these beaches a few days later, I couldn't count the dead. A billion animals had died. All this, worth one-fifth of one point along the current gold standard for measuring our collective emotional response to a warming world. One wonders if it's the scale or the observer that is miscalibrated.

I called Kathy Selvage because someone had told me she could explain what it was like to lose a mountain. Near her home in Wise County, Virginia, in the name of the seams underneath, summits are lopped off and King Coal shoves his arms in up to the elbows. Selvage, sixty-six, has watched these ridges atrophy for decades. "It's a dead zone," she tells me. "You're not only being harmed physically. You're being harmed spiritually. That intimacy with the mountain is gone."

Mountaintop removal, it's called: a type of surface mining. Mountaintop removal—one of those euphemisms that whisper of the passive voice. Who, exactly, removes the mountains? From whom are they being removed? What else are we removing?

Selvage knows about the intimacy of the mountains because she has spent a life living in their shadows. Around the turn of the millennium, she founded Southern Appalachian Mountain Stewards, a community organization dedicated to halting the destruction caused by surface mining and improving the quality of life in the region. Most recently, she was a caretaker for her mother, who passed away in 2015. "In many ways I grieve for her," she says. "But I associate the loss of the mountains with grief too. That probably sounds rather strange, but I think of them as one and the same. I mourn them both."

Before our first conversation, Selvage sends me a photograph of herself in which she's standing in front of a NO TRESPASSING sign. Wearing high-waisted blue jeans and an orange T-shirt, both faded, she is gripping the sign with her right hand. Behind her: stone, cinder blocks, sand, cement. There is a steely stare at the camera. When I ask her about the photo, she explains: "The homeplace where I grew up is under all that rubble." "Homeplace" is one word. It had been bulldozed.

There are two things that come up in nearly every conversation I have with people from or familiar with mountaintop-removal towns. The first is a strong, immutable sense of place—a connection to the region and to the mountains that this journotourist from pancake-flat

Minnesota has little chance of grasping. Selvage's stories of her child-hood are of barefoot stream wandering, of crawdads and minnows, of fruit trees at the neighbors'. They are stories of trips to the moun-taintops, berry picking; of owls, pheasants, and quails. "To have all that demolished, taken away, geographically eradicated? It is one of the most disturbing things. It was the death of the mountain," Sel-vage says.

The second thing that comes up in these conversations is the use of rape metaphors for what the coal companies have done to the land.

Over the past decade, mental health professionals and epidemi-ologists working in Central Appalachia have begun to translate these feelings of spiritual violation into cold, hard research: statistics and case studies bearing academia's wax seal of approval; studies that say the violation is real. In 2012 and 2013, for example, two independent teams of researchers published a pair of articles in *Ecopsychology*, an academic journal that covers the niche field wedged between its namesakes. The studies painted a new picture of mountaintop re-moval's public health crisis: not only as one of asthma and cancer and birth defects, but also as one of substance use disorder, anxiety, in-somnia, and clinical depression. Compare a mountaintop-removal county to another coal-mining county in the region. Even when ad-justing for educational status, for socioeconomic outcomes—anything that could potentially skew the results—symptom rates of clinical depression are significantly higher in mountaintop-removal mining areas than in other coal-mining areas in the region. This is to say nothing of the elevated rates compared to those in non-mining towns. Whatever the reason, people living in mountaintop-removal counties are more than one and a half times as likely to exhibit at least moder-ate depressive symptoms, compared to those in non-mining counties.

In a non-mining town in the region, we might expect around ten people out of one hundred to have moderate to severe depression. For mountaintop-removal areas, that number is closer to seventeen out of one hundred. Two million people live in the area.

Of course, in many ways, studies like these confirm what residents of Central Appalachia already know: The death of a mountain cuts and scars the people living at its foot. Scars are just easier to see when they're on the lungs.

"The more traditional biological issues make more sense," offers Michael Hendryx, an author on the 2013 *Ecopsychology* study. Hendryx was director of the West Virginia Rural Health Research Center from 2008 to 2013 and is now a professor of applied health science at Indiana University Bloomington. As he points out, it's easier to imagine how something like coal dust might be associated with something like asthma. Depression, on the other hand, is less tangible—and not something easily pinned to the aluminum and silica in the air. But it's not about air pollution or water quality. "For depression," says Hendryx, "it seems to be more about the destruction of the environment. It's that psychological impact of watching mountains blown up and roads destroyed and communities wither away and jobs disappear and politicians lie to you every day about what they're doing."

Hendryx's idea isn't a new one. Around the same time that Kathy Selvage was founding Southern Appalachian Mountain Stewards, an Australian philosopher named Glenn Albrecht was coining a term that would help describe these feelings of environmental loss and place-based distress. For Albrecht, "solastalgia" was to be contrasted with nostalgia: It was the pain and yearning that stem not from the passing of better times, but from an altered environment. Albrecht developed the concept—whose name derives from "solace," "desolation," and the Latin root for "pain"—after witnessing the psychological impacts of open-pit mining on local communities in New South Wales. Where surface mines go, solastalgia follows.

Paige Cordial, a clinical psychologist in Southwest Virginia and lead author on the other *Ecopsychology* study, explains the idea like this: "It's homesickness, but you haven't left home. You're homesick because the landscape has changed around you."

Since its inception, the term has also been applied to survivors of natural disasters, as well as to Ghanaian farmers and Inuit communities faced with changing climates and landscapes. Cordial, who has run the Central Appalachian circuit—originally from West Virginia, she went to undergrad in Kentucky and grad school in Southwest Virginia—has also documented anecdotes suggestive of post-traumatic stress disorder in mountaintop-removal towns. Frequent blasts and unpredictable floods could, for some, be "acute and severe enough" to lead to PTSD, she writes. She cites the story of West Virginian anti-mountaintop-removal activist Bo Webb, who has also spoken out against the psychological effects of surface mining. "I know first hand the mental effects of shell-shock," said Webb in a 2011 interview. "I witnessed it and experienced it in Vietnam. It is evident to me that people living beneath and near this terror are experiencing much the same."

In their publications, Hendryx and Cordial both zeroed in on solastalgia to help explain the apparent disproportionate spikes in depression rates in surface-mining towns. "You can't really go home," says Cordial—and that displacement begins to fester. Her doctoral dissertation focused on the psychological impacts of mountaintop-removal mining. She recalls the words of one of her interviewees: "Everything just seems so strange. Like a moonscape. Everything that used to be so beautiful: Now it's just like walking on the moon."

In his book *The Great Derangement*, Amitav Ghosh argues that the unpredictability of climate change—its corresponding rejection of bourgeois regularity, its uncanniness, its reminder of powers outside our direct control—offers cause for considering that "nonhuman forces have the ability to intervene directly in human thought." As we've seen countless times in this book, the claim is not as fantastical as it sounds. Climate anxiety; degradation-spurred depression. These can be dulled and subtle forces, but they are everywhere. In Washington State, during fire season, the mountains are fuzzed out and we feel glum; in West Virginia, the mountains are decapitated in service of a mining operation, and major depressive symptomology increases one-and-a-half-fold. The twinge you feel, searching for the mountain that had been embedded in your understanding of the world? It's the effect of neurochemical changes prompted by a lack of recognition—or, perhaps, as Ghosh would argue, from a recognition of the nonhuman forces we've so readily discounted. People may have removed the mountains, but the mountains themselves were propping up something important within you. You've changed, just a little. And this pang is nothing next to the literal unmooring and rewriting of identity that occurs when, say, the climatological effects on nearby river conditions force you to abandon the practice of subsistence fishing your family has relied upon for centuries.

In the halls of neuroscience, identity formation is generally understood to encompass three distinct processes: recognizing the self, establishing independence, and, perhaps paradoxically, affiliating oneself with groups and practices. Behind these processes is a tangle of neuronal pathways we have yet to completely decode—but the neuroanatomy isn't so important here. What is important is the realization that identity formation is no solitary voyage. It is a journey deeply entwined with the social landscape around us, and such entanglement should remind us of the *relational* nature of the self. And on this front, neuroscience does have something to say. Consider the humble mirror neuron.

Our first encounter with mirror neurons harks back to a lab in Parma, Italy, in the 1990s. Neuroscientists studying macaque monkeys stumbled upon a set of cells in the frontal cortex that fired not only when a given monkey performed an action, like reaching for a peanut, but also when it simply observed another monkey performing the same action. It was as though these neurons were holding up a mirror to the observed action, reflecting it back within the observer's own brain.

Since then, evidence for similar neurons in humans has been accumulating, and while the exact nature and extent of our mirror-neuron system remain hot topics of investigation, the implications of these findings have sparked widespread fascination. Intricate networks of mirror neurons appear to be responsive to the actions and emotions of others. The notion of empathy comes to mind. Presumably, these cells reflect and help shape our social interactions. (Recall the phenomenon of neural coupling, in which storytelling synchronizes brain activity between teller and listener.)

At their core, mirror neurons propose a simple yet elegant solution to a profound problem: *How do we understand the actions, intentions, and feelings of others?* Traditionally, this process has been thought to require complex cognitive gymnastics—a rapid-fire series of inferences and deductions. Mirror neurons, however, suggest a more direct route: embodied simulation. We understand others because their actions, their emotions, and their pain resonate within our own neural circuits—as if we were performing the action, feeling the emotion, or experiencing the pain ourselves.

Is it such a stretch to imagine we might mirror our natural environment too?

We know that identities—these collections of practices and behaviors we wrap around ourselves—are not fixed but continually shaped and reshaped by the terrain of our experiences and relationships. We know they are not just a product of our individual neurobiology, but also of the intricate interplay between our neural landscapes and the social, cultural, and environmental ecologies we inhabit.

We know that in the face of a charred forest, a guillotined mountain, a parched field that once reliably nourished—identity slips. To blot out a landscape is to interrupt a rhythm, and it is rhythm through which an identity remains static. If neuroscience can claim any fundamental truths, one is that learning and memory are rooted in newness and surprise. Slot machines activate the neural pathways of addiction precisely because we can't predict which pull garners the jackpot. Little surprises along the way keep us coming back. As Karl Friston might argue, we most easily remember the improbable—and when it comes, we change. A missing mountain is no little surprise. It is shocking.

Our understanding of regularity, which Ghosh details as having been born in the Enlightenment and buoyed by the bourgeois lifestyle, represents a way of thinking about the world that's new, relatively speaking. He argues that the idea of a life governed by slow, plodding predictability—as understood by the modern novel and the nine-to-five—hinges on a set of assumptions about the world that the world ultimately and continually disproves. When we build coastal properties, we operate under the assumption that they will not be struck by tsunamis. When we overleverage ourselves with an all-too-easily-attained McMortgage, we assume the housing market will continue to climb.

Events like Superstorm Sandy and the financial crisis and the coronavirus and mountaintop removal scoff at us. And then they shape us. Biologically, no, it's not yet entirely clear how the death of a mountain and the birth of a moonscape wind up translating into clinical depression. Certainly, it's infinitely more complicated than the previous sentence suggests. And it would be irresponsible to suggest that all or even the majority of those living in mountaintop-removal towns, for example, respond this way. Appalachia is already a region painted in broad strokes: as what photographer Roger May described in 2015 as "the last bastion of America that's sort of generalized, lumped together, and made fun of." Adding "depressed" to the "celebratory hillbilly" stereotype does nobody any favors. But for those who are susceptible, epidemiological evidence does appear to suggest a relationship between mountaintop-removal mining and poor mental health. Not every kid in a hot school performs worse on a test; but more than usual do.

In October 2015, a team of researchers at Emory University and the Cooperative Institute for Climate and Satellites published a paper spelling out the logic of environmental degradation and mental health outcomes. The causal web is one of "rupture of place bonds, culture change and loss, and altered community and family dynamics," all of which, the public health scientists demonstrate, rigorous research has linked to anxiety and depression. How might an altered relationship with one's environment bear on mental health? Through identity loss, they argue.

Indeed, the Emory researchers' logic is the kind that rings true to

Selvage. "You're not just destroying the land," she says. "You're destroying the people. You're not just eradicating the mountain; you're eradicating a community and its people's history along with it."

Previous work by Hendryx, the Indiana University Bloomington researcher, has shown that 70 percent of rural Appalachian counties have a shortage of mental health care workers—a percentage significantly higher than that of rural counties elsewhere in the same states. A 2009 nationwide study also concluded that many Central Appalachian counties were among those with the highest unmet need for mental health professionals. Seeking out care in coal country is simply harder than it should be.

That's assuming you actually want to seek the care out in the first place. But that would imply that the broader rural stigma against mental health treatment doesn't exist. It does; and it doesn't make the picture any prettier. Talking about mental health isn't something one tends to do in the area. "Let's say you go to a counselor," says Cordial. "Everyone knows your car." That's a problem, she says, because many people still think that only "crazy" people see counselors or psychiatrists.

In practice, all these swirling structural clouds make the state of the region's mental health care something of a threefold perfect storm. Economic and environmental stressors lend themselves to higher incidences of substance use, anxiety, and depression; a dearth of mental health professionals leaves the need unmet; and the stigma against mental health treatment, combined with what has often been described as a certain fatalism of Central Appalachia, means that plenty of people aren't interested in seeking out the few available resources in the first place. Compared to healthy people living in rural areas, people in poor mental health are also more likely not to be insured. And while Obama blaming is still big in the region, coal blaming is not.

"Coal is still big here, but it's going down fast," says Cordial. "That makes people even more hesitant to talk about its negative effects." When an industry's history is as embedded in a region's as coal's is in this one, turning on the industry when it's in decline is a little like not calling for help when your stepdad goes into cardiac arrest. Truly conceptualizing the grip that King Coal has on its Central Appalachian subjects is outside the scope of this chapter, but it is real and it is a stranglehold.

"You have this extractive industry that takes what's valuable and removes it," says Hendryx. "Almost all of the wealth leaves. It's the definition of a resource curse." When I first called him up to ask him about his research, Hendryx immediately noted that he didn't begin his work with an axe to grind. That's a salient point, because the researcher has come under heavy fire from the coal industry for his publications. In 2015, one company spent considerable resources—and filed a lawsuit—to force West Virginia University, where Hendryx was previously based, to hand over thousands of documents related to his research. On the phone, he is a dispassionate shade of incredulous. Of the mining and public health connection, he says, "It was just what I found."

Says Cordial, "And don't forget that coal companies have given a lot of money to universities in the region."

There are many interests aligned in trying to convince you to forget these sentiments and these stories and these datasets, to embrace a collective amnesia. It's like that joke that says when you spin the country record backward, the wife and the dog come back. Of course, it never works that way. The mountains are still dead, the sound all scrambled no matter which way you spin it.

Selvage recalls her mother seeking solace on her family's porch over the years, where she would sit each morning with her Bible, alternating between reading and looking out over the landscape. "She should have had that right," she says. But then the coal trucks came, and with them their dust and the departure of the mountaintops. That solace left with the peaks.

And maybe there is nothing to be done about the loss of the mountains. It took 400 million years to build them—they're not coming back overnight. Depression, too, is something that is managed, not cured. The erasure of stigma takes decades; poverty doesn't just evaporate. "Progress" isn't a word on many lips here. It's hard to stand on the shoulders of giants when we keep cutting them off.

Originally trained as an auditor, Selvage is all business when she talks about the future. "I have never seen a place that has cried out for a strategic plan more than the coalfields of Appalachia," she says. "Because if we are depressed and unhappy and stagnated and don't know what to do, then we are not going to appreciate progress. At least until we have something by which to measure it."

For the rural mental health sector, though, progress is something that can be measured. Progress looks like the expansion of mobile health services. It looks like the expansion of telemental health services, which in turn requires the expansion of broadband access. School-based programs. The development of consistent reimbursement systems for telehealth. Keeping homegrown mental health professionals like Paige Cordial in the region. Maybe there is nothing to be done about the loss of the mountains, but maybe there are new mountains to rise.

"Hope" is a complicated word for Selvage, but when she feels it, it's tucked away in the promise of knowledge creation; in recognition of the problem, in regional planning, in entrepreneurship, in reversing the brain drain she sees in the region. "I do believe that there's hope for Appalachia, but I don't believe it can transform itself," she says.

"We have fired this nation. We have lit this nation. What now can the nation do for us?"

On the other side of the country, in December 2017, when Jennifer Atkinson flew home for the holidays, she was met with a blackened Earth. Thick smoke snuffed out the stars she had grown up sleeping under; the land beneath the darkened heavens was unrecognizable. By the middle of that month, the Thomas Fire had become the fourth-largest fire on record in California, consuming 242,000 acres. Fires weren't supposed to burn like this, at least not this late in the year. "To actually breathe in these toxins, to know they're going into your lungs and your body, there's a sense of assault, or a kind of invasion, that is just deeply personal," Atkinson reflected recently.

When California burned again, in 2018 and 2020, prior years would look like a grim dress rehearsal. In 2020 alone, more than 4 million acres burned: the state's worst fire season in modern history. Falling ash would rain down like locusts. Irrigated vineyards—historically firebreaks; in recent years sucked dry by the leech of climate change—fell like spears before assault rifles.

The violence and intimacy of the wildfire was something Atkinson would bring back to Seattle after her visit home. She was set to teach a new course: an undergraduate seminar at the University of

Washington's Bothell campus called "Environmental Grief and Anxiety: Building Hope in the Age of Climate Consequences." She'd advertised for the course with flyers reading DEPRESSED ABOUT CLIMATE CHANGE? When dozens of students enrolled in the seminar, she knew she had struck a chord.

The chord has something to do with the social function of this type of education, of grappling publicly and collectively with all this anxiety and depression. Which means it also has to do with identity. And with pain.

In the experience of near death, our sense of time dilates. Survivors of car crashes, lightning strikes, the pummel of waterfalls: They almost uniformly report a sense of clarity and calm characterizing the would-be final moments. Anna Jane Joyner mentioned as much in her telling of her near shipwreck on the Tasman Sea. The neuroscientist David Eagleman has theorized that the brain records more sensory information in traumatic experiences, and that when we look back on those elaborately recorded moments, our biological clock appears to have slowed down. Peering over the brink, our eyes open wider.

What, then, in the experience of a protracted death? In the long premortem of global warming, climate anxiety is something of the opposite of clarity. It is a swirling and a rug pull; a gnawing and a mourning; a quickening and an aberration of time. It is an "elegy for a country's seasons," as Zadie Smith wrote. It is imprecise. It is pain.

Pain is interesting, neurologically, because it is true alchemy. Your peripheral and central nervous systems work in concert to convert a stimulus—say, a cut on your thumb—into a mental representation. These systems can dial back and forth the suffering you feel; perceptual grades of pain don't necessarily follow degrees of injury. Sure, your brain converts photons bouncing off a knife into a mental representation of that object. But it doesn't invent the knife. The pain, it invents.

And yet the point of pain—its neurological raison d'être—is that you can't ignore it. Pain's "very nature demands attention," writes the anesthesiologist Abdul-Ghaaliq Lalkhen. It swivels the mind's eye toward the body's mistake, as perhaps climate worry swivels it toward humanity's. Don't forget the magic, though: Pain is interpretation. It takes ages, electrically speaking, to convert cellular damage to histamine and bradykinin, to transmit news of these releases to the dorsal horn of the spinal cord, to pass the message along up through the medulla to the thalamus, to form some semblance of representation of injury in the somatosensory cortex—all before the understanding and the suffering begin. You process an injury and attach meaning and emotions to it; your brain drapes the sensation of pain over your perceptive window.

The fact that pain is characterized by *sharing meaning* implies that it indeed has a social function. Lalkhen again: "It is only when an individual processes the injury in terms of attaching meaning to it that the expressions of suffering and pain are exhibited. In this way, pain is a form of communication. It allows other people to see what the injury means to us as individuals, and it has survival value, in that our expression of pain will produce empathy and assistance." Pain, in its raw intensity, is a powerful communicator. It shatters any illusion of solitude and beckons others into our sphere of experience. In the animal kingdom, we see examples of as much in the alarm calls of meerkats, the danger signals of bees, or the empathetic grooming of primates. In humans, as Lalkhen suggests, an expression of pain or a cry of distress draws immediate attention, triggering reactions of empathy, concern, and assistance. Is the same not true of our expressions of emotional pain? Shared empathy can offer a mutual acknowledgment of the struggles that underscore our common humanity.

Further, pain is a reminder that our brains exist in bodies and that our bodies exist in space. We need to eat. We need to exercise. We need to treat our bodies like bodies.

Consider a hand recoiling from a flame. The motion itself isn't a detached response to a painful stimulus. It is a movement of self-preservation: an embodied narrative of danger and escape. The hand doesn't just transmit the sensation of pain—it enacts it, inscribing in its movement a certain story of suffering. Our experience of pain, too, is deeply entwined with our emotions, thoughts, and memories. A painful experience can summon memories of past pains, evoke emotions of fear or distress, and trigger thoughts of harm or threat. Pain is a complex, multidimensional experience, a cloth woven of sensation, emotion, cognition, and memory. When we remember our bodies as bodies, through the lens of embodied cognition, pain reveals itself as an intricate part of our existence: a testament to the deep interconnectedness of mind, body, self, and world.

Perhaps no greater phenomenon captures this interconnectedness than climate migration, itself often a withdrawal from flame. Climate migrants begin a journey born of necessity, not of choice. This is not a tranquil meandering. It is a journey made in desperation, etched in the stark language of parched fields, scorched forests, and swallowed coastlines. It is a narrative of displacement, where the protagonists are not explorers or adventurers, but ordinary people, in developed and developing countries alike, driven from their homes by an antagonist as mighty as it is elusive: a changing climate.

In developing countries, often situated on the front lines of climate change, this phenomenon is especially stark. Take Bangladesh, a land cradled by rivers and exposed to the Bay of Bengal. Here, rising sea levels and repeated floods gnaw away at arable land, swallow homes, and salt the fields, rendering them barren. People have begun to find themselves dispossessed, their livelihoods drained away by the very rivers that once nurtured them. Compelled by the ruthless advance of water, they set out in search of new beginnings in overcrowded cities or foreign lands.

In developed countries, too, climate migration unfolds, albeit with a different texture. Consider the wildfires that have ravaged parts of California—like the Thomas Fire that shocked Jennifer Atkinson—reducing dream homes to ashes, scarring the landscape and the people who inhabit it. Here, affluence may offer more options for relocation and the resources to rebuild life elsewhere, but it does not confer immunity from displacement.

In cities like Miami, the reality of rising sea levels is seeping into everyday life. Water regularly rises through storm drains, floods roads, and infiltrates homes. As neighborhoods become periodically aquatic, properties on Miami's higher grounds, often in historically low-income neighborhoods, have become increasingly attractive. This newfound allure drives up prices and drives out long-term residents, who find themselves unable to keep up with the skyrocketing cost of living. The neighborhood of Little Haiti provides a stark case in point. Here, the elevation is just a few feet higher than in the rest of Miami, but those few feet represent a lifeboat in a sea-rise scenario. Accordingly, developers are moving in, luxury apartments are sprouting up, and longtime residents are being squeezed out—their culture and community dispersed in the face of climate-driven economic forces.

To understand climate migration is to confront the searing physical realities of our warming world. It is to trace the footsteps of families leaving low ground and drought-stricken farms. It is to listen to the quiet despair of fishing communities watching coral nurseries bleach in acidifying oceans; to witness the silent retreat of islanders before the advancing tides, the sea swallowing ancestral lands as easily as it laps at a shell. Climate migration paints a picture of the Anthropocene, where human fingerprints mark not only the atmosphere and the oceans, but also the migratory patterns of entire communities. The greenhouse gases we've sent aloft on the wings of industry return to us as extreme weather events, forcing millions from their homes in an echo of our own actions: a planetary-scale feedback loop that spares no one.

There is a qualitatively distinct pain that accompanies these climate-induced migrations. It is not merely the strain of movement or the distress of leaving familiar surrounds. It is a dissonance between our innate longing for home—for continuity—and the imperatives of survival in a changing world. Every family hearth extinguished fractures the continuity of lived experience and intergenerational memory, echoing a loss as profound as the extinction of a species. It is a loss so obviously embodied that it hardly seems worth pointing out. It is.

What to do with all that embodiment? How to bear it? In the face of climate anxiety, of degradation-fueled depression, of the chronic stress of climate migration, we would do well to recall Karl Friston's idea that the self-reinforcing model—the organism that gets stuck—is that which fails to interact with its environment as a means of coping. The pain comes from somewhere. Can we focus our attention elsewhere?

Consider, for example, a remedy offered by the forest: the practice of shinrin-yoku, or "forest bathing." Born in Japan, this therapy invites us to immerse ourselves in the sensory wilderness of the woods: to listen to whispering leaves, to inhale the fragrance of earth and bark, to gaze upon the shifting patterns of light and shadow. By forming an intimate bond with the forest, the logic goes, we not only find momentary respite from our worries but also kindle a deeper understanding of our place in the web of life. Such connection and understanding might, in turn, empower us to face our anxieties with clarity and purpose.

Equally, we might find solace in the practice of "deep time" contemplation. Thinking on the scale of the geological past—and into a deep, unknowable future—we can learn to see ourselves as part of a vast temporal spectrum. This perspective can dilute the intensity of our immediate anxiety, situating our climate concerns within a broader context of geological change and the resilience of life. Such a temporal shift does not diminish the urgency of addressing climate change, but proponents argue that it can help modulate our emotional response, swapping paralyzing fear for constructive concern.

We can also turn to citizen science. We can engage in small acts of data collection and observation. These contributions, minor as they may seem individually, add up to create a significant body of knowledge that can inform climate action. In becoming a part of the solution, even in such a small way, we may find our anxiety giving way to a sense of agency—that rare thing we're trying to reclaim.

And again there is the simple act of storytelling. In sharing our fears and hopes about the changing climate, we create a narrative: a collective dream of the future. Through this act, there is opportunity to morph our individual anxieties into shared resolve, our solitary worry stitched into the quilt of collective action. In telling our stories, we may find that the shadow of climate anxiety gives way, even a bit, to a dawn of renewed resilience and shared determination.

It's hard. Anxiety and depression and chronic stress sit deep within us. As parts of our minds, they are embodied too.

As we've seen, when the shadows of stress loom, our amygdala stirs, sending forth a neural call to arms. This alarm triggers the hypothalamic-pituitary-adrenal axis, setting in motion a cascade of hormonal responses that culminate in the release of cortisol, the body's primary stress hormone. In the midst of fleeting stressors, cortisol is a critical ally, marshaling resources, sharpening our focus, and priming us for action. But when stress persists, the continuous deluge of cortisol can leave the landscape of the brain altered. The hippocampus is particularly vulnerable. Bathed in chronic cortisol, the region can atrophy, its neural branches withering. Such an effect can lead, unsurprisingly, to impairments in memory and learning.

Similarly, the prefrontal cortex, our master planner, is susceptible to the eroding influence of chronic stress. Impacted by sustained cortisol exposure, our ability to think clearly, make decisions, and regulate our emotions can be compromised, as if a fog has descended upon the paths of reasoning. In the context of climate change, these neurological effects can be exacerbated. The upheaval of seeing one's home change, or, indeed, leaving it altogether; the uncertainty of seeking refuge; the struggle of adapting to new environments—it can release relentless cortisol storms.

Certainly we harbor the capacity for resilience and adaptation. Neuroplasticity, the brain's ability to rewire and reshape itself, offers one seed of hope. In some sense, the phenomenon *enables* resilience, giving us the capacity to learn from hardship, to adapt to stressors, and to change our perspective and response to challenges. In turn, by fostering resilience—through social support, mindfulness practices, or therapeutic interventions—we can help steer the brain's adaptability toward growth and recovery. Chronic stress and climate anxiety, while formidable, are not invincible adversaries. In understanding their neural echoes and promoting strategies of resilience, we might still navigate these turbulent waters.

But it does take remembering that your body is a body. It is more than your mind.

On a recent trip to Oslo, at sunset, I climb a hill to the bridge and vantage point depicted in *The Scream*. A small plaque confirms I am in the right spot. Because it is the middle of winter and I am the only fool to be standing in the cold on the side of the road, I am disinhibited, which is to say that I feel comfortable turning around and making the face from the painting. It seems like the right thing to do, as I'm sure it has felt like the right thing to do for many an Oslo tourist. Turning and looking back down over the city, I half expect to see the sky twist and redden. It does not.

I think of Annie Dillard:

The mind wants to live forever, or to learn a very good reason why not. The mind wants the world to return its love, or its awareness; the mind wants to know all the world, and all eternity, and God. The mind's sidekick, however, will settle for two eggs over easy. The dear, stupid body is as easily satisfied as a spaniel. And, incredibly, the simple spaniel can lure the brawling mind to its dish. It is everlastingly funny that the proud, metaphysically ambitious, clamoring mind will hush if you give it an egg.

I descend the hill for a very late lunch.

9

THE GRAMMAR OF EARTH

Language is the only homeland.
—Czesław Miłosz, first utterance unknown

I had an Anthropocene ox on my Holocene tongue.
—Robert Macfarlane, *Underland*, 2019

GEASSI (SUMMER)

In Russian, the color blue is divided into two distinct categories—голубой (goluboy), denoting light blue, and синий (siniy), encompassing darker varieties. Speakers of most other languages do not distinguish between these hues in this manner. In English, for example, голубой and синий are merely shades. They do not rise to the linguistic pedestal of full "colors." It is a curious little quirk in the broad evolutionary history of human language—and color perception, for that matter—that one linguistic group should differentiate between these flavors of blue. And so in 2007, a team of MIT researchers, led by the linguist Lera Boroditsky, decided to investigate the effects of this bifurcation on cognition. Did the additional linguistic granularity, as represented by more words for "blue," affect speakers' psychology? What about their ability to navigate the world? Language offers such an intimate bond between speaker and environment. Might it also offer a psychological bridge between perception and reality? Boroditsky and her colleagues asked volunteers—speakers of different languages—to distinguish between various hues across the blue spectrum and to do so as quickly as possible.

The linguistic partition between shades appeared to enhance Russian speakers' abilities to identify them. When presented with a range of blue hues, they showed faster and more accurate blue discrimination compared to speakers of languages lacking the голубой–синий distinction. In particular, the effects were most extreme for difficult tasks, in which the hues in question were perceptually nearest to one another. It was like the Russian speakers had better color vision—like they had an extra variety of rod or cone that allowed them to more effectively distinguish between subtle stimuli. But they didn't. They just had an extra word.

This type of effect isn't limited to Russian. The Indigenous Namibian Himba, for instance, have *no* distinct word for blue in their linguistic palette. In experiments, they have been observed to struggle in distinguishing between what English speakers would identify as blue and green—even though they can discern the subtleties in the latter's spectrum far beyond an average English speaker's ability. Boroditsky's research has also explored how the presence or absence of certain linguistic categories can affect memory and other cognitive processes. Russian speakers, for example, are ultimately better at remembering shades of blue than shades of green. Language can tint the lenses through which we view the world, determining the hues we perceive.

The research challenges a long-held idea: namely, that color perception is a universal and invariant phenomenon. Boroditsky and others' work—on what has been deemed linguistic relativity—underscores the idea that language plays a vital role in shaping perception. Like a prism refracting light, language colors our environment. It colors our cognitive processes. And it does so in more than just shades of blue.

ČAKČA–GEASSI (AUTUMN–SUMMER)

Mandana Seyfeddinipur removes a cat from her office chair and sits in its spot.

"You are sitting in front of a Whorfian," she says. This is another way of naming herself as a linguistic relativist, like Boroditsky.

The moniker comes from one Benjamin Lee Whorf, an early-twentieth-century linguist who studied the interplay between language, cognition, and culture. Whorf's seminal contribution to linguistic theory came in his exploration of the idea that language is not merely a medium of communication, but a potent force that shapes our perception of the world. Studying the Hopi language, with its intricate verb tenses and grammatical structures, he noted that within the language was embedded a distinct temporal framework that emphasized the cyclical and contextual nature of events. There was no emphasis on measurements like hours or minutes; instead, the language paired the significance of events with larger cyclical patterns like seasons, natural phenomena, and the rhythms of agricultural life. And importantly, it appeared this temporal orientation was embedded within the fabric of Hopi linguistic structure. Whorf would go as far as to hypothesize that it helped guide the Hopi people's conception of time itself. Time wasn't an entity to be quantified and segmented, but rather a fluid and interconnected flow.

In this manner, language can facilitate a deep appreciation for and fluency with the interplay of past, present, and future. It blurs the boundaries between them, altering how Hopi speakers *perceive* time. They aren't outliers, either. Unlike English speakers, who spatialize time from left to right, Aboriginal Pormpuraawans, for example, organize time from east to west—that is, the progression of time is in accordance with the sun's movement. This linguistic frame influences spatial cognition.

Just as Boroditsky's research on Russian would challenge the universality of color in the twenty-first century, Whorf's research on the Hopi language stood up to the prevailing notion of time as a linear progression. By extension, it encouraged a broader understanding of temporal diversity across cultures. And more generally, Whorf's work suggested a profound influence of language on cognition. While some of his ideas have since been refined and debated, this notion—of language shaping thought—laid the foundation for many current investigations in linguistic anthropology. When we discuss linguistic relativity, we're discussing Whorfianism.

Seyfeddinipur grew up in Germany to Persian parents from northern Iran. The first point she makes about the contextual links between language and reality relate to this background: "I speak Persian to the degree of a fourteen-year-old," she says. "Because that's all I needed in the home. When you're at home, you talk about what you're going to eat, when you're going to go to bed, and when you're going to go on holiday." Today, Seyfeddinipur speaks other languages, frequently with more prowess than a teenager. She is concerned with language shift and language loss, and she directs the Endangered Languages Documentation Programme and the Endangered Languages Archive at the Berlin-Brandenburg Academy of Sciences and Humanities and the University of London. There are currently about seven thousand languages spoken globally. Roughly half are expected to survive the next hundred years.

Languages fade for many reasons. One significant factor stems from the outside influence of dominant or politically powerful languages—English among them. Globalization, colonization, and cultural assimilation have played pivotal roles in suppressing Indigenous and minority languages. When speakers of these marginalized languages are forced to adopt dominant tongues, intergenerational transmission of Indigenous languages can be disrupted, leading to their gradual erosion and erasure. Stigmatization can accelerate language loss, and the power of the state presses inward. Policies that prioritize certain languages as official or prestigious while marginalizing others can result in language shift and endangerment. Economic opportunities and urbanization, too, can draw younger generations away from their homes and from other minority-language speakers. In pursuit of education or employment, people shift to languages that potentially offer greater access and mobility.

But there are also climatic and biodiversity factors at play. Small linguistic communities on islands and coastlines—or in areas in which changing temperature and rainfall patterns threaten agriculture and aquaculture—are subject to climate displacement. Relocation and forced dispersal of people necessarily lead to language splintering, as communities come into contact with speakers of other languages and fewer speakers of their own. And as climate refugees settle in new communities, it's often advantageous to speak the dominant language of the area, for the same reason any other migrant might work toward a similar linguistic shift.

In other words, as our climate changes, it doesn't only alter landscapes; it also erodes the linguistic diversity that blossoms from the human experience of these very environments. The world is a matrix of languages, each contributing to the richness of human life and expression. Climate change, in all its tumultuous forms, accelerates the degradation of this richness. As seasons become unpredictable, the words that once charted the course of the year lose their bearings. As species die out, the words for them risk slipping into obscurity, their linguistic niche disappearing like the dodo or the passenger pigeon.

In all cases, though, Seyfeddinipur emphasizes, we're talking about people and culture. "In our metaphor around the linguistics problem, we talk about languages dying or falling silent," she says. "And the interesting thing about the metaphor is we are abstracting away that languages don't die, you know? They don't fall silent. They don't do anything. It's people that die. It's people that are actors in their own rights, and we're taking away the activism in or the agency in people shifting." Climatic factors are one among many that move people away from mother tongues.

That is the obvious component of the interplay here. It is the nuance, though, that Seyfeddinipur finds most compelling. In her linguistic relativism, language helps us construct reality, but it also helps us capture the natural statistics of our environment. As Karl Friston might say, it helps us navigate our econiche. This is what's happening when Seyfeddinipur speaks of learning just enough Persian in the home. Her language matches the structure of her environment. And this emphasis on environmental structure captures a distinct component of climate-fueled language loss. One doesn't need to be physically displaced to experience a language shift. "The environment changes practices," she says. "People change, and they stop doing certain things, and they no longer use the relevant vocabulary. So something goes stale."

RÁGAT (RUTTING)

Language gives boundary to thought. It is a container that holds emotion and idea—a leaky container, but a container nonetheless. Outside of language, where does a warm drink end and a hot drink begin? It is through linguistic precision that we can give structure to the messiness of the world—and through linguistic creativity that we can play with that structure.

Think about snow. There are blizzards and squalls, flurries, the unholy "wintry mix." When snow settles, it can take the form of powder or slush or sleet or crust or something called graupel, a kind of soft hail characterized by little snow pellets. One can shovel drifts. One can shred the gnar. English: a curious, complicated language. Gaze at its majesty. Let's be generous and say it has twenty words for snow.

Northern Sámi has 318.

Speakers of the language, who largely populate the Sápmi region stretching across the northern reaches of contemporary Scandinavia, have reason to relate intimately to the snow. Sápmi is frequently cold. It gets a lot of precipitation—and employment in fishing and reindeer herding has necessitated a nimbleness and precision with language that allows fishers and herders to communicate environmental conditions accordingly.

But as the climate changes—rendering whole swathes of Sámi snow types extinct and their corresponding names detached from reality—vocabulary suffers in kind. As winters shorten and temperatures fluctuate, as snow and ice recede and transform, the essence of these words—their referents in the natural world—are altered or made uncertain. In this way, the impact of climate change on the Northern Sámi language is not just symbolic but strikingly literal. Specific words for "snow" or the various stages of reindeer antler development represent concepts intricately linked with survival and tradition. As climate change grips the Arctic, the balance is broken. Winters arrive late and end early, and snows that once were reliable now come unpredictably. The intricate calendar of nature's milestones, etched over centuries into the vocabulary and grammar of the Northern Sámi language, is being scrambled by the disordering effect of climate change. Unpredictable weather patterns, warmer winters, and shifting landscapes are transforming the life that the language has, for so long, intricately described and cherished.

Climate change doesn't merely transform the physical world; it also reconfigures its cultural and linguistic landscapes. Is a language a mountaintop? As languages are lost, so, too, are the histories they carry.

Seyfeddinipur frequently makes this argument with respect to languages' abilities to encode traditional knowledge. "If we monoculturize our wheat everywhere—because we have the genetic power to do that—and make it resilient to heat and cold," she argues, "what happens when a new parasite comes to town? The entire crop: Everything is gone. But if we have ten different varieties, while eight might die, at least two will survive." She thinks language is like this. We cannot monoculturize it. No one language can do it all.

VUOSTTAŠ MUOHTA (FIRST SNOW)

I fly from Oslo to Alta, in the Norwegian Arctic, where the snow is coming down hard and the buses are canceled. I learn the latter fact from a Norwegian transit app that I need to iteratively screenshot and translate via search engine for the message to sink in. There are no taxis in front of the small airport's taxi stand, and I begin to ponder a two-and-a-half-hour walk through a blizzard. Just as I am tightening the laces on my boots, a cab pulls up. I approach the driver, perhaps a bit desperately.

As we drive the twenty minutes northeast to the small town of Rafsbotn, where I will tend to a friend's farmhouse and pick up a Volvo for the rest of my trip, my taxi driver complains about the snow. It's not that there's more of it, he says, but that it's frequently wetter. He tells me he often has to pull over and wipe off his headlights when driving in storms like this, that it never used to be this way. He does indeed do this twice during the short drive.

In the brain, language representation has the look of a winding Norwegian highway. It twists across the cerebral landscape, touching dozens of neural regions and frequently making pit stops to dust off its headlights. There are the primary areas dedicated to language processing, like Broca's area, situated in the frontal cortex, which plays a pivotal role in the production of language—orchestrating the motor movements required for speech. There is Wernicke's area, in the temporal cortex, intricately involved in language comprehension. It allows us to decipher meaning from the words we hear or read. These language-specific regions form the core of what is known as the classic language network.

But across this neural terrain is a sprawling network of interconnected regions that contribute to various aspects of language processing. Beyond the classic language areas is the angular gyrus, for example, residing at the crossroads of the parietal, temporal, and occipital lobes. It helps integrate visual and linguistic information, playing a role in reading and semantic processing. There is the superior temporal gyrus, alongside Wernicke's area, which contributes specifically to the comprehension of spoken language. The language network comprises white-matter tracts, such as the arcuate fasciculus, which serves as a pathway connecting Broca's and Wernicke's areas, facilitating the flow of information during language processing.

Which is another way of saying that as language changes, much of your brain must change in kind. Studies of multilingual speakers illustrate the brain's ability to flexibly engage and switch among different language systems, showcasing the dynamic nature of language representation. This flexibility translates to other tasks. (Consider the case of Russian blues.) The brain is plastic. When we lose language, this flexibility deteriorates.

In the echoing halls of memory, too, language plays a decisive role. As events unfold before our eyes, our hippocampus is busy inscribing these happenings into the parchment of our past. But how we recall these events—how we make sense of them—is intricately tied to the narratives our language allows us to construct. Our autobiographical memories are therefore not perfect replicas of our past, but narratives shaped by the storytelling tools our language provides. As we've already seen, our understanding of abstract concepts, such as time or moral judgments, is also deeply intertwined with the linguistic scaffolding our brain erects. Metaphors, the poetic devices of language, might shape how we perceive such abstractions. The metaphor of "time as space," for instance, where we talk of a "long" time or a "short" time—as though time were a physical entity—influences our internal representation of an otherwise elusive concept.

Language can offer insight into a new dimension. Imagine being handed a black-and-white photograph of a toucan. You could understand much about this thing: its shape and size; perhaps you would guess it could fly. But if you were then handed a color photograph of the same bird, you would understand something else about it. You probably couldn't unsee the color if you went back to the black-and-white photo. Vocabulary is like this: New words give access to a precision of experience that old words do not. In both cases, you're looking at a photograph of a toucan. But in one, you're missing something important. Absence of language to describe the details of the Earth—details harvested from experience and connection—is like missing the family member you never knew existed.

In this light, the climate crisis and the language crisis are two facets of the same problem: a certain disconnect from the natural world. They reinforce each other, creating a feedback loop of loss. Language, tethered somewhat as it is to a particular environment, cannot survive the uprooting of its speakers, their diaspora and dislocation, without transformation or—in some cases—extinction. And each language lost is a unique prism of human perception gone.

SKÁBMA (DARK TIME)

The more we understand about the psychology of language, the more we see it is linked to our landscapes. Like a river shaping the contours of a valley, the climate, terrain, and ecological niche in which a community thrives can influence the phonetic and lexical elements of its language. The languages of the highlands, for instance, have been found to contain more consonants, possibly because the thin air at such altitudes makes it easier to produce such sounds. A striking example comes from the language of the Quechua people, nestled high in the Andes, which is rich in ejective consonants, sounds produced with an intensive burst of air.

The *ecology* of our environment also shapes our language. Communities living in close proximity to rich biodiversity often have a more nuanced lexicon for plants, animals, and their interactions. An example here is the language spoken by the Matsés tribe in the Amazon, which includes a remarkably detailed vocabulary for the myriad species of flora and fauna that populate their world. The Kuna people of Panama, in their dialogue with the sea, have multiple words for the different stages of a turtle's life, mirroring their intricate existence in their coastal habitat.

Psychology also points to the social and cultural environment as a potent shaper of language. Social conventions, cultural beliefs, and communal practices all seep into a language's veins, giving it its unique hue. The Aboriginal language of Guugu Yimithirr, for example, which uses compass directions for spatial orientations rather than those relative to the body (such as "left" and "right"), is a profound reflection of a culture deeply attuned to the geography of its land.

Neuroscience affirms these influences. The brain, like clay on the potter's wheel, is sculpted by experience and exposure. Neuroplasticity ensures that our neurons and their connections are not static, but dynamically shaped by our interactions with the environment. The area of the brain most heavily involved in spatial cognition, the hippocampus, is more developed in London cab drivers, who regularly navigate the city's complex streets, as well as in speakers of languages like Guugu Yimithirr, who rely on cardinal directions.

Multilingualism in particular shapes the brain in distinctive ways. Neuroscientific studies suggest that managing multiple languages enhances cognitive flexibility. It calls on the brain to nimbly switch between language systems, bolstering executive functions like attention, problem solving, and task switching. This workout for the brain can foster what is known as cognitive reserve, offering a robustness that ultimately may delay the onset of diseases like Alzheimer's and other dementias. Each language we learn creates new pathways in our brain, enriching its intricate neural network. Brain imaging studies show that multilingual people often have a more substantial gray-matter plasticity in certain areas associated with language processing and cognition.

But there is a beauty to multilingualism that transcends its neuroscientific value. As we've seen, each language represents a unique way of perceiving and describing the world. Each offers a different perspective: a unique narrative in humanity's ongoing dialogue with the world around us. To speak multiple languages is to move fluidly between different landscapes. It is to carry within oneself, in essence, a reflection of biodiversity. Thus, the value of multilingualism is at least twofold. In the ever-shifting landscape of the brain, it cultivates resilience and flexibility, offering tangible neurocognitive benefits. But perhaps equally important, it enriches our understanding of the world and our place within it, allowing us to better appreciate the complexity and diversity of the human experience, to better remember it, to tell its story.

Anouschka Foltz, a linguist at the University of Graz, in Austria, is interested in what happens when languages collide. Multilingualism, for Foltz, is the gold standard of human experience. "Globally, handling one language is the exception," she tells me. "I'm interested in how the brain manages this collision." And broadly speaking, it manages it quite flexibly.

Fundamentally, her work, like Seyfeddinipur's, pushes up against the use of Standard American English and Standard British English as measures of competence. "When we talk about language competence, we use standardized tests and we measure how well people can function" using one of these languages. But Foltz thinks this question is the wrong one. We shouldn't understand the health of language in any person as encoding how many people they can successfully communicate with, she argues. Instead, she thinks true competence is all about how flexible we are in our language processing. "Often what linguists do is compare everyone with the ideal native speaker, which is the monolingual native speaker of a language," she says. But in the contemporary maelstrom of dialects, that speaker probably doesn't even really exist. "So the question is: Are these the skills we want to measure?"

In the realm of machine learning, where algorithms learn to predict the world's behavior from patterns in data, a peculiar paradox of perfection arises: the problem known as overfitting. The phenomenon unfolds when an algorithm, in its zealous quest to map the intricacies of its training data, ends up lost in a labyrinth of its own precision.

Imagine you're trying to predict the future behavior of a tree's leaves based on years of meticulous observations. You've watched their ebb and flow, their bud and wilt, across seasons and storms. Now, armed with this data, you set out to create a model. In a world without overfitting, you might notice general patterns. Leaves bud in spring, thrive in summer, and fall in autumn, and branches are bare in winter. This model, simple yet effective, captures the tree's cyclical dance with the seasons. It's a model that understands the tree but isn't distracted by the whims of individual leaves.

In the grip of overfitting, your model would aim for perfection. It would account for every rustle, every tremor, every leaf that fell too early in autumn or clung stubbornly through winter. It would create a complex map so intricately tailored to past observations that it would predict each quirk with astounding accuracy. But herein lies the problem: When faced with a new season, your model would stumble. It would search for those individual leaves, those specific tremors, those exact rustles. Nature, of course, in its beautiful chaos, wouldn't reproduce those quirks precisely. Your model, in its pursuit of perfection, would falter, losing the forest for the trees. Overfitting is a cautionary tale about the cost of overzealous precision. It reminds us that the goal of any modeling effort is not to echo the past with perfect fidelity, but to learn from it to predict the future. It's about finding that delicate balance where the model is complex enough to learn from data but not so complex that it loses sight of the underlying patterns that will help it flexibly navigate the future.

Anouschka Foltz's work on multilingualism raises the stakes of language loss. It says monolingualism—the global trend we're implicitly following, as languages are lost—amounts to a flattening of experience. By leaning too heavily in the direction of, say, English, we collectively lose whole modes of relating to the world. As a species, we become less adaptive. We are overfitting.

DÁLVI (WINTER)

It is January in Guovdageaidnu, which means the sun only peeks above the horizon for two or three hours each day. It is blisteringly cold and punishingly dark, but when that little orb traces its spoon-edge arc over the landscape, the sky has the look of a perpetual sunset. Twenty hours by car north of Oslo on a good day, well into the Arctic Circle, Guovdageaidnu is the coldest town in mainland Norway. (It is known as Kautokeino in Norwegian.) It is the reindeer-herding capital of Sápmi, which is to say that it is also the de facto reindeer-herding capital of the world. The rhythm of Guovdageaidnu is that of the Arctic itself—seasons of eternal sunlight and perpetual night; a cycle of extremes. In summer, the land stretches under the ceaseless gaze of the midnight sun, its expanse embroidered with the blooms of Arctic flowers and the murmur of rivers. In winter, a polar night descends, transforming the landscape into an ethereal world of snow and starlight, frequently illuminated by the otherworldly dance of the northern lights.

Jonna Utsi meets me at the Pit Stop Café, where the top review on Tripadvisor is: "I assure you, it's open." Utsi has been in the mountains for several days, and it has been difficult to connect, but when we sit down to our coffee, he is fully present, and his smile comes easy.

Utsi's family's reindeer herd is thousands strong. In the extremes of the Arctic, the reindeer are the truest experts, he says, guided by a primal intuition that senses the subtle shifts in wind and light, the faint rustle of lichen crunching beneath the snow, the unseen dangers lurking beyond the horizon. They move with the seasons, following ancient migratory routes that crisscross the Nordic wilderness. Walking in their footsteps are Sámi herders like Utsi—custodians of a knowledge passed down through generations. Theirs is a dominion not of force but of understanding and respect. He explains it as a bond forged in the crucible of Arctic survival, where the line between life and death is as thin as the ice beneath one's feet. Each herder knows their reindeer, understands the subtle nuances of their behavior, the language of their movements, the melody of their calls.

I wanted to talk to Utsi because I wanted to understand what it might be like to watch a language crumble before one's eyes. By this point, I had read plenty of news reports about Northern Sámi words for snow disappearing as a function of climate change. I admit it seemed a little too easy, a little too romantic. Was that really how it worked?

Without question, and without prompting, he noted that the climate was changing, and it was making his life vastly more difficult. "When the snow gets warm and the ice gets warm, it gets a little bit wetter," he began to explain. "It gets heavier. When it freezes, it gets harder. If it's just one day, it's not a problem. But in the last ten years, it has happened almost every time. The winter doesn't come before January. Instead, it rains, and the rain freezes as plates."

This is bad news for the reindeer, who have to break through the plates to graze. "The real problem is that the reindeer start to look for new places to graze," says Utsi. "Even when they find a new area, they have to hit the ice plates with their legs. They can't get to the food. They start to starve." The first winter in memory that played host to this problem was 1997. Maybe a few years here and there in the following decade—and then most of the past ten.

When life is harder for the reindeer, it's harder for the herders. They end up needing more fuel for their snowmobiles. They need to buy supplementary food for their reindeer and drive it out to their herds—thousands of pounds at a time. They need to stay with their herds longer, tending to the animals and ensuring they get enough to eat. They stay in the mountains overnight, often for days on end.

It wasn't always this way, and for Utsi, this is where language loss enters the equation. Growing up, he explains, winter was always a time to learn the ways of the world. The reindeer were fine where they were, and his father could return to their lavvu, their home tent, in the evenings. It was at night, beside the warmth of the firelight, that the family would share stories about the day, about the trade, about the past. "Even if they were telling a very stupid story, all the time they were using the language," he says. "They would tell how it was, up in the mountains. How the snow was that day. They used the vocabulary of the language." The home, at night, was where learning took place—including learning of Northern Sámi and its manifold words for "snow."

But now, as they have to travel greater distances across rougher terrains, it doesn't always make sense for herders like Utsi to return home at night. He isn't always telling these stories to his children. "That, I think, is the biggest change in language I see," he says. Forced by the climate to remain in the mountains, he simply doesn't have enough time to teach his children—to share his stories—in the manner he'd prefer. "We have lost a lot of our knowledge," he says. "It's very bad."

I'm reminded of something Anouschka Foltz told me about the importance of individual words: "The vocabulary of a language is the thing that encodes the knowledge. It's not the grammar that encodes the knowledge." Words are what we use to navigate the vast tundra of shared human understanding. They are more than just strings of letters or sounds; they are containers of meaning, holding within their structures the sum of our cumulative knowledge about the world. In this sense, our vocabulary is both an encyclopedia and a compass—encoding what we know of the world, while also guiding us through the terrain of human thought. As vocabulary fades, we're left out in the cold, tapping on ice plates for lichen that may not be there. The weight of nature always requires we press harder.

DÁLVEGUOVDIL (MIDWINTER)

"If I didn't become a social linguist, I would have become something like you," Annika Pasanen tells me. "Or an environmental activist like my sister. That's what I'd planned to do." We are sitting in Dr. Pasanen's office at Sámi Allaskuvla, the Sámi University of Applied Sciences in Guovdageaidnu. And, indeed, she didn't become an activist. In 1995, Pasanen moved to Inari, a Sápmi region with its own dialect of Sámi and a healthy language-revitalization program. Wanting to learn the local language, she began engaging with the program. And as she learned Inari Sámi, she became interested in language revitalization more generally.

"Language formulates our identity—our belongingness," she tells me. "That's why global language loss is a tragedy to humankind. The more language we lose, the poorer our culture." She is somewhat of a Whorfian, but not to the extent of Seyfeddinipur. She thinks language influences our reality—no question—but she wants to know more about why and how. What she is certain about is that language is never just language: "It carries the whole history of a particular group."

These are the stakes of climate-driven language loss. Far past blooming lakes and wildfires, deeper than mountaintop removal, more insidious than a heat wave, language loss cuts both our present reality and our past. The words we use, Pasanen says, the rhythm of our speech, the metaphors we reach for—they all serve to paint the canvas of our individual and collective identities. Every language holds within it a world of wisdom and experience. Each word is a distillation of human understanding, an acknowledgment of a shared reality. When a child learns the language of their ancestors, they are not just learning words: They are inheriting a worldview, an ethos, a way of being in the world. In this way, just like the environment, language shapes our identity, defining the contours of our thought, feeling, and understanding.

Yet, says Pasanen, identity, too, shapes language. The words we use, the dialects we choose, the stories we tell: They change and shift as we navigate the winding paths of our lives. Our individual and collective experiences etch themselves onto the language we use, in turn shaping the linguistic environment for future generations. There are words for "snow" in Northern Sámi that a Hawaiian might never know, just as there are Hawaiian words for subtle aspects of the sea that may remain forever foreign to speakers of Northern Sámi. Our languages are both map and mirror.

Acknowledging and prioritizing as much is a political decision, she says. In some countries, legislation explicitly bans minority or Indigenous languages. "And in others, they are not officially forbidden, but in practice, it is impossible to speak them in public places because of the negative attitudes and feedback you will get when speaking anything other than the dominant language." For Pasanen, this is just as much a ban on identity as explicitly disallowing migration from a given region. Additionally, the political and economic pressures associated with climate change often lead to further marginalization of Indigenous communities. Exploitation of the Arctic for its resources, driven in part by easier access resulting from melting ice, often displaces Sámi communities, severing the ties connecting the people, their language, and the land.

"And parents are very vulnerable," she says. "In the face of negative attitudes, all parents of the world want the best for their children, and they make linguistic choices accordingly. Parents want to choose a language that they think will make a better life for their children." Pasanen's research illustrates the manners in which the political environment shapes language, stealthily and indirectly, just like the climate.

We can read Jonna Utsi's reflection on language loss in a similar tone. He's not actively choosing to teach his children English—they learn enough of it in school—but without the opportunity to give them as much Northern Sámi as he'd prefer, he worries the language will fall to the wayside completely. He speaks of politics in the same manner he speaks of climate change. "It gets in the way," he says. And as we've seen countless times, the climate continues to creep in. It doesn't much care for politics, either.

GIĐĐA (SPRING)

Utsi's and others' tellings of language loss in the Arctic have a feel of realism that I couldn't find in academic papers. I am reminded of Ryan Wallace from the CDC telling me that he was concerned about the climate's effects on political instability—and the resulting knock-on effects on rabies surveillance, prevention, and treatment. The indirectness of an effect makes it no less legitimate. Bank shots still count in billiards.

"I could never figure out how some of the things posited in theories could be psychologically real," says Foltz when I tell her about my first conversation with Utsi. But the realism makes sense to her. There's no magic here, just the weight of nature.

In linguistics, reckoning with this weight implies pushing back.

Language revitalization is akin to regrowth after a wildfire. It is the cultivation of linguistic seeds that have lain dormant, their vitality preserved beneath the ashes of disuse or suppression. Revitalization is a celebration of linguistic diversity—the acknowledgment of a language's unique reflection of human experience, its wisdom and worldview, its ability to push back against overfitting. Each language is an ecological niche within the wider landscape of human cognition and culture. As in any ecosystem, diversity within the linguistic landscape fosters resilience and richness of thought. When a language is revitalized, we regain a part of this diversity, rewilding the cognitive and cultural landscape.

Further, according to Foltz, the value of language revitalization lies not only in the preservation of cultural heritage and the enrichment of cognitive diversity. There is also a healing aspect to it. For communities who have had their languages suppressed or endangered, revitalization can represent a reclamation of identity, a strengthening of communal bonds, and a defiant act of resilience against historical trauma.

We have examples of successful language revitalization. One comes from Hawai'i, where the native Hawaiian language was on the brink of being lost. By the 1980s, English-language assimilation policies had left fewer than fifty native speakers under the age of eighteen. It was a language in winter, muffled by a thick blanket of snow. But the Hawaiian language soon began to stir. In 1983, committed speakers launched the Hawaiian Language Revitalization Project. The effort ultimately sparked the establishment of Hawaiian-language immersion preschools, or Pūnana Leo, the nest of voices. The voices grew in number and strength, echoing through the establishment of K–12 immersion schools, then reaching higher education with degrees offered in the Hawaiian language. Today, Hawaiian once again resonates through the islands' valleys, a testament to the resilience of cultural heritage and the power of organizing.

Another story of revitalization can be found in Wales. In the twentieth century, the Welsh language was dwindling. Through concerted efforts, however, including comprehensive Welsh-language education and the creation of a Welsh-language television channel, the language has since seen a significant resurgence. Today, Welsh holds an official status in Wales, and more than a quarter of the population are speakers—bards of their own ancient tongue. Consider, too, the vast landscapes of New Zealand, where the Māori language, or te reo Māori, was also in danger of falling silent. The Māori community, like their Polynesian cousins in Hawai'i, rallied. Kōhanga Reo, or "language nest" preschools, were established, alongside Kura Kaupapa Māori, primary schools where Māori is the medium of instruction.

Māori is now an official language in New Zealand, with many young people gladly embracing their linguistic heritage. Much of Annika Pasanen's work on Inari Sámi takes place in a similar Inari language nest, across the border from Sámi Allaskuvla.

Not that the work is easy. Dominant languages, often backed by political and societal power structures, can always crowd out smaller, Indigenous languages. Overcoming this dynamic requires not just the teaching of words and grammar but a broader shift in attitudes: a recapturing of values that recognize the worth of every language.

It can also simply be challenging to find fluent speakers and comprehensive teaching materials for endangered languages. Sometimes, the elderly are the only fluent speakers left, making the task of transferring knowledge to younger generations a race against time. Moreover, revitalization isn't merely about teaching words and grammar, but about reviving an entire ecosystem of cultural practices, stories, and ways of seeing the world—a holistic endeavor that is as complex and intricate as the interdependencies in any natural ecosystem. Revitalization is about nurturing an understanding of the language as it exists in a web of relationships—with the land, the community, and the broader culture.

Rarely linear, the path to language revitalization is frequently a long, winding road. Language revival requires time, patience, and nurturing to take root and flourish. It demands the creation of spaces where the language can be lived, breathed, and experienced in daily life—an environment conducive to its growth and preservation. This is no small task in a fast-paced, globalized world, where change is often driven by speed, efficiency, and uniformity.

But in Foltz's and Pasanen's estimates, the political realism here, captured in part by Utsi's story, helps reinvigorate language revitalization to a certain degree. If language loss were merely a matter of melting ice caps, what could they do? Mitigating climate change is hardly the goal of linguistics. But politics, parental decision-making: You don't actually need to refreeze the ice caps to effect change in these areas. Not all environmental activists look like Pasanen's sister. Many more look like Pasanen herself.

That is the grand lesson for all of us who don't live in the Arctic or work as social linguists on language revitalization. Our contributions to climate solutions—to the effects of environmental degradation on brain health—will come from within: from our own strengths and predilections and opportunities. These solutions, those that we design in the interest of our mental and emotional resilience, don't require us to become "activists" in the sense of carrying a sign in a march. They require us to actively notice the change as it creeps. They require us to become activists in our own lives.

GUOTTET (CALVING)

Máhtte Sikku Valio noses his Peugeot around the darkness of Guovdageaidnu. From the passenger seat, I listen to the crunch and slip of snow under the tires, and to a faint whistle of wind, and to Máhtte's curving voice as he points out the landmarks. He's a journalist, a friend of a friend, who has agreed to show me around town. When I note that he is brave to drive a French car through so much snow, he jokes that it is in his blood, because he is a little wild, because his mother was "born in the wild"—because she was born in a lavvu. "Possibly I need a Volvo next, though," he offers, as the tires briefly don't quite catch. We turn off the road into a construction site.

Lit up with floodlights in the afternoon black is a sloping, angular project. I can't quite understand its shape, but I can see that one of its faces appears to run off the roof like a snowdrift. Máhtte throws the Peugeot into park and tells me the completed building will eventually house the Sámi National Theater, as well as a new Sámi high school. It is high contemporary architecture, designed by Snøhetta, the same firm responsible for the Norwegian National Opera and Ballet. The building's name is "Čoarvemátta," the Northern Sámi word for the thickest part of a reindeer's antlers: the point just where they depart the skull and begin to branch. Today, most schooling in Guovdageaidnu is conducted in Northern Sámi, and the same will be true here. Having reached that daring post-middle-school age of specialization, students will be able to study reindeer husbandry at the complex, if they so choose. Many will.

We swing by Sámi Allaskuvla for a coffee and sink into a booth in the university's cafeteria. It is late, and we are the only people here. I realize I have yet to learn what the name of the town means. In

Northern Sámi, says Máhtte, it is something like "the middle of the road." Halfway; a crossroads. "In Sámi, Guovdageaidnu *means* something," he emphasizes. "But the translation or Norwegification of the word to 'Kautokeino'—it doesn't make sense. It doesn't mean anything. It is destroying language." He sounds tired. Máhtte is coming off of a flu, and before that, a long recovery from brain cancer. Our ramble about Guovdageaidnu is the longest he has walked and talked in a while, and it is time to head home for dinner. He will return to his family; I will eat my eighth evening veggie burger at the Thon Hotel, and then I will make my way back north to the farmhouse, away from the middle of the road.

Does a word like "Kautokeino" destroy language? Certainly it masks it. The word is powerless without its adoption, though: It can't destroy meaning unless someone *uses* it. But part of Máhtte's argument goes beyond meaning. Language is more than vocabulary, he argues. It is also culture. To mask culture—perhaps especially to mask it with meaninglessness—*that* is true erasure. When he says Norwegification is destroying language, he isn't talking about signposts and definitions. Language is a glue. And when Norwegian doesn't translate the full meaning of Northern Sámi, it doesn't just break it down: It also separates the people it holds together. It is no coincidence that so much of climate change operates in this manner.

Later, I find an architectural rendering of the Sámi high school site Máhtte had driven me past. The unfinished sloping face I'd noticed in person, it turns out, will form an asymmetric curve along the length of the site's southwest end—a continuous transformation from roof to facade to ground. There is a blending here, but the site is stable in its fluidity. It does not disappear into the landscape. The building meets nature, yet it also stands distinct. I think: It feels like an activist in its own life; that there is an echo of the grand lesson here.

From above, the school is something of a three-point star with the points lopped off. It is a stocky, partite thing—a cellular division—a marker of growth, maybe. It occurs to me that there is a kind of reaching-out action to be read in the shape: a reaching from the mind, a reaching through the skull to the warming hillside beyond. It is a grounded, yearning reach; an anchored split; indeed a kind of čoarvemátta.

COUNTERBALANCE

I am at dinner with friends in the basement of Clube de Fado in Lisbon, one of the oldest and, I am made to understand, most venerated fado establishments in Portugal. The fado form is an old genre of music and a poignant relaying of human experience. Rooted in the cultural soil of the country and the Portuguese language, fado braids together melancholy melodies, aching lyrics, and heartfelt performance. The name translates to "fate" or "destiny."

Having emerged in the early nineteenth century in Lisbon, fado tends most frequently to reflect the yearnings, joys, sorrows, and aspirations of the local, in particular, capturing the Portuguese essence of saudade—a complex sentiment that encompasses a deep longing, nostalgia, and bittersweetness. It reflects the relation between person and place. At the heart of fado lies the fadista: the singer who becomes a conduit for raw emotion. Accompanied by a classical or Portuguese guitar, the fadista's verses speak of lost love, longing for home, the trials of life, and the resilience of the spirit. In 2011, fado was inscribed on UNESCO's Representative List of the Intangible Cultural Heritage of Humanity, recognizing its significance as an emblematic expression of Portuguese culture and its ability to resonate with the universal human experience.

As we'd walked to dinner, Quinn, one of my friends, had offered to our party the English word "petrichor": the smell that arises from soil or asphalt after the rain; the essence of the land, released with water. Lisbon's petrichor today is oily and of jacaranda. And as we now observe in the basement of Clube de Fado, along with the scent of honeyed, purple flowers, a few of the town's ghosts appear to have been released. It feels that way, anyway: that something is afoot. That there is movement with us. That the spirits have followed us indoors. It is warm here, out of the rain. It smells like history.

An hour into dinner, the fadista enters the club to no applause. The lights are lowered, a hush descends, and the fadista smiles. She has noted the ghosts in the room. She closes her eyes, leans back, and begins to sing.

The leaning is significant. The spirits press. Each of the three fadistas over the five-hour dinner service will strike the same pose tonight: eyes closed, ghosts confronted, backed into a corner, shoulders resting against a stone pillar. But it is an easy lean. The pillar cradles the fadista, as it has cradled hundreds of fado singers before. Quinn notices that if you look closely at the point at which her head and shoulders touch the stone, you can make out a polished, worn smoothness that just forms the gentlest cup. The fadista leans into a bed of sorts.

The weight of nature presses, but we get to press back. A depression is a state, but it is also a refuge, a place from which to come forth.

feel like I'm in full empathy with our planet," Dezaraye Bagalayos, from the California Central Valley, had written to me. It wasn't an expression of peace. Dezaraye's planetary empathy is rooted in pain: the aches and grimaces of a world grieving not just the loss of species and an unspoiled troposphere, but also the loss of the way time once passed and the seasons once progressed. It is the sorrow of glacial ice bearing water for the last time; seas unable to hold their vapors close; the erasure of whole fluvial languages once carved into riverbeds. Of fires bigger than the last, of a world losing the words that describe it. Earth is hurting, and empathy means hurting with it. The grief is heavy.

Full empathy. I remember, a few years ago, reading the obituary of a lawyer who had set himself on fire in Brooklyn. "My early death by fossil fuel reflects what we are doing to ourselves," he wrote in an email to reporters beforehand. His name was David. I had stared at his photograph on my computer, searching for something—what?—in his expression. We will never know his albatross, not the way he did. But we do know David had felt this planet. He'd said so, right there in that note. Maybe his was full empathy; the fullest.

I'm not sure. There are other ways to think about the word "full." A central effort here has been to illuminate the connections— foreboding and terrifying at times—between the human brain and the turbulent world wrapped around it. The fullness of it. But this book is not in its fullness a story of despair. Rather, for me, it is one of reciprocity and desire. Of: How can we sing a depression into a refuge?

I think my desire speaks less to adaptation to future climate change than it does to adapting to the present in which we've found ourselves. Certainly, we *can* adapt to a new, climate-changed world. But I believe doing so begins with the world that's already here. We have needs now.

We can understand our mental and emotional intimacy with this nature as a portal to understanding a world that is more than our own—and as a means by which we can more deeply relate to one another. When we acknowledge these forces pushing and pulling at us, when we really *feel* the contours of the winds marionetting our heartstrings, the acts of climate change are no longer abstract and alien. We are coming to know Earth. We are hearing it breathe, and we are breathing with it. Grief is only ever a sign that you've loved something enough to care that it's changing. The weight of nature is an anchor, and it binds us to this world.

In this way, our planetary connection—this momentous, miraculous communication between home and tenant—is not merely something to be adapted away. Rather, we can sit with our grief and pain and anxieties, together, to feel less alone. We can reframe them as connective tissue and shared experience. We can deepen our language for our experiences and share our stories with each other. To live in full empathy with the planet is to be bound up in the lives of one another and in those of our children and forebears.

This empathy is intergenerational. It is cross-species, intercontinental, and geologic. And I believe it can offer the channel by which we finally respond to the climate crisis. The most cynical of policymakers and scientists argue that action won't come on climate change until enough of us have felt the crisis come crashing into our lives. Perhaps. But we're already under the crashing. Recognizing and welcoming our heavy planetary bonds may offer a means of realizing a societal response to the climate crisis—a manner in which we can build the connections that are necessary for the collective action that protest, policymaking, and accountability require. Anchored, steadied by grief, we can act.

On a recent April evening, my partner and I sat on a bench off a quiet street in north Seattle, watching boats and paddleboarders navigate a thin stretch of the Lake Washington Ship Canal. We'd brought wine and pizza: vegan, with little nubbins of fake chicken and red onion and some healthy dollops of barbecue sauce. Someone on the other side of the canal skipped a rock with such proficiency that Allison screamed in joy. The sun was thinking about setting.

I am somewhat of a romantic, if that hasn't been already made abundantly clear. Ally is, among other things, an environmental epidemiologist; the science brings a kind of wry and calculating reductionism to her worldview. So when I looked out over the water and vocally marveled at the chain of improbable events that had brought the two of us together, this is what she said: *Sure, but that's true of literally everything that has ever occurred. Everything that happens follows from the infinitesimally improbable chain of events that preceded it.*

A nonchalant bite of barbecue chik'n. And then: *That doesn't take the magic out of it. It means we were meant to be together, by order of the universe, no less. Determinism doesn't negate the magic of the journey.*

Okay, she didn't actually say that very last line about determinism. But that is what I heard. Much of the subtext of this book speaks to a kind of soft deterministic view of the world, one that points to the manners in which the environment can strip of us of some imagined agency—but also how that lack of free will ultimately fails to excuse us from acting with compassion.

I think what Ally's point adds to this picture is a reminder that the line between push and pull is always fuzzier than it seems. That the journey is the point, because the thing that happens next is going to happen. And that doesn't take the magic out of it.

Here is what happened next for us. First, you should know that immediately before us was a tree, and I am here to tell you I can't remember what kind of tree it was. It had leaves. That's the quality of nature writing you get in this epilogue.

What happened was: The earth rotated just a little more, and the angle of the sun's rays quickly suggested pirates' gold, and all of a sudden the tree before us was lit up like December, the holidays, ten thousand shiny coins for leaves, all glinting something dappled. It was as if someone had flicked on a whole neighborhood of lights. And then my already dazzled noggin lit up with something too, a thought along the lines of: *Isn't it stunning that we have a word to describe the specific quality of light seen filtered through a tree canopy?*

I suppose in that moment, we must have been dappled together. I believe in many moments, people are dappled. Sometimes, we see it. Much of the time, though, we are focused on the trees in front of us. Spending too much time on the trees can obscure the forest, yes, but the act can also neglect our own experience. I am trying to notice what happens to myself in these moments—trying to catch a glimpse of the sun on my skin too. I am trying to remember that a rainbow, that visual dazzler every bit as stunning as tree dappling, doesn't make any sense without the context of an eye to sense it and a brain to perceive it. It seems worth paying attention when the world performs a filtering; when we are struck with rays in a new manner; when we are touched by something weightless.

ACKNOWLEDGMENTS

I can't begin an acknowledgments section without the names Larry Weissman and Sascha Alper: double-trouble agents who took a relatively unwarranted chance on a first-time author who didn't really know how to write a book proposal. Larry and Sascha, thank you for your vote of confidence, your professional guidance, and your healthy edits on that initial document. I hope I've done you proud here with the rest of it. This whole section of the book could be about you, but I'll save us all the embarrassment.

Stephen Morrow and, later, John Parsley at Dutton shepherded this project on its journey from rambling manifesto to reasonably legible treatise. I'm so grateful for your enthusiasm for the book, your wise editorial advice, and, most importantly, your faith in my vision for the project. Laura Stickney at Allen Lane: Thank you for your insights, your thoughtfulness, and all those dang book recommendations—all of which have made me a better thinker and a better writer. I'm lucky to have landed with a stable of editors by whom I've felt prioritized and understood. Horror stories from the publishing industry tell me that it doesn't always go that way. Thank you, too, to Grace Layer, Ella Kurki, Fahad Al-Amoudi, Hannah Poole, Diamond Bridges, Annabel Huxley, and Julie Woon. I'm deeply grateful for your time and attention.

This book wouldn't exist without Grist and its early-career fellowship for environmental journalists. Some of the first scratchings of mine on this topic were provoked by conversations in 2015 with my first editor, Andrew Simon, who saw possibility where I did not. Lisa Hymas, Katharine Wroth, and Ted Alvarez each offered valuable editorial insights and helped me begin to develop my voice—something I'd never particularly considered before meeting them that year. Additional Grist-adjacent thanks are due to Teresa Chin and Nikhil Swaminathan, who later in my career wholeheartedly supported my commitment to this project and helped me craft a relationship with the magazine that allowed me to actually, you know, write the book while still holding down a part-time job. Thank you for your patience and your investment in me.

Thanks to Jason Mark, who was perhaps the first person to encourage me to get serious about writing a book proposal, as well as to Sue Halpern, Bill McKibben, Nate Johnson, Danielle Svetcov, Rebecca Bright, Atul Gawande, and Siddhartha Mukherjee, all of whom offered sage advice on how to navigate the famously opaque publishing industry. Further early encouragement to explore the ideas that wound up in this book came from late-night conversations with Niklas Allamand Dib, Julian Gewirtz, Thomas Hale, Moss Amer, Brian Troyer, Amelia Bates, Mubeen Shakir, Alex Michie, and Tim Hwang. Thank you for these discussions and for the ideas you've no doubt seen reflected here. Thanks to Gregg Colburn, who tolerated my split attention between this book project and ours. Thank you to Lauren Gravitz, whose reporting on active forgetting informed my own, and to Erin Welsh and Erin Allmann Updyke, whose reporting on *Naegleria fowleri* introduced me to Sandra Gompf's story.

I have benefited from various fellowships, residencies, and writing stipends over the years of working on this project. My deep honor and gratitude are due to Artist Trust, the City of Seattle's Office of Arts & Culture, Two Dot Schoolhouse Studios, and Mesa Refuge. Zach Frimmel, Ruthie Tomlinson, John Tomlinson, Kamala Tully, and Eirinie Carson deserve special thanks on that front, as do my Mesa housemates Irma Herrera and Rita Cameron Wedding, dinner discussions with whom often influenced the writing herein.

Thank you to early readers of some of these bits and bobs, among them Lowell Wyse, Naveena Sadasivam, John Thomason, Eve Andrews, Jonathan Pease, Ryan Batic, and Neal Morton, all of whom offered clarifying and imaginative feedback. I learned and grew from conversations on these topics with Britt Wray, Dan Sherrell, Burcin Ikiz, Angie Michaiel, Leslie Davenport, Rachel Malena-Chan, Lydia Carrick, Beihua Page, Emelio DiSabato, Mackenzie Brown, Carson Cornbrooks, Whitney Henry-Lester, Andrew Palmer, Cristina Friday, Jeff Alvarez, Ana Machado, Ashley Davies, Andy Staudinger, Nancy Rottle, Paul Olson, Tresa Smith, and Jake Bittle. Thank you for your time and your wisdom. A million thanks to Liza Birnbaum and her Yearlong cohort, especially my buddies Ruthie and Daryl— and my patient friend Karen Taylor—who read more drafts of some of these chapters than feels reasonable. Liza, you are a profoundly talented teacher. Thank you.

Anneka Olson, thank you for your deep conversations on these topics and your encouragement over the years. This project probably wouldn't exist without you—not in its present form, anyway—and for that, I owe you my deepest thanks. I'm grateful for your editorial eye, your sharp insights, your good humor, and your realism. Thank you for pushing me and for holding me accountable.

Edward Boyda and Svea Vikander offered lodging and introductions and friendship in the Arctic. They also let me borrow their farmhouse and their Volvo, the latter of which I succeeded in not running off the E45 on the way to Guovdageaidnu. Thank you for trusting me. You are dangerously benevolent.

It would be impossible to list every source that has informed the writing of this book, but I have to extend my deep gratitude to Manny Aparicio, John Cassani, Michael Reed, James Metcalf, Dezaraye Bagalayos, Kathy Selvage, Jennifer Atkinson, Jonna Utsi, and Karl Friston. There is a book to be written about each of you, but I hope the inclusion of some of your ideas and experiences here might suffice for the moment. Thank you.

Darby Minow Smith—my collaborator, professor, editor, dear friend, and co-operator of the Hungry Horse Rootin' Tootin' Residency for Hungry Artists—thank you for showing me how to pay attention, how to be goofy, how to follow my nose; perhaps, in short, how to write. Let me know what's next, coach. Chips ahoy.

To my mom, Bonnie: Thank you for teaching me how to observe the world with the eyes of an artist. I hope you see the sacrifices you made to raise me right have borne some kind of fruit in this project. Thank you for gifting me a life that afforded the opportunity to write a book. Thank you for finding Brad Aldern, who undoubtedly resonates through these pages, as well.

A hearty thank-you to other family members: among them Ben, Buffy, Chris, Roxanne, Noah, Alisa, Noreen, Bob, Maxine, Adam, and Susan, all of whom were forced to tolerate me thinking through some of the initial ideas here at a family reunion in 2016 . . . and no doubt again, several more times, over the years. Thank you to Jared,

Mary, Jackie, Grayson, Natalie, and Jimmy for doing the same in a different part of the country (and occasionally, a different part of the world). Thank you to the Dog Ranch and its various tentacles, all mentioned above, which opened its various doors when I needed a home and a writing desk. Thank you to Charles Lenth, to whom I wish I could show the thing we spent many hours discussing over Maker's Mark.

Allison. I confess I struggle to communicate the gratitude I feel for your support of and your intellectual contributions to this project. You've invited me to think about the world in manners that sometimes feel completely new. Thank you for refocusing my attention on compassion, empathy, and presence. Thank you for authoring paragraph-long text messages on receptor densities when I find an illegible toxicology paper and reach out in a panic. Thank you for introducing me to vital cheerleaders Sophie and Toca. We all joke you're something of a vampire queen. Thank you for making me undead.

NOTES AND FURTHER READING

PROLOGUE: TENSION

1   *"like I don't fully exist"*: Dezaraye Bagalayos, email correspondence with the author, September 23, 2019. Unless otherwise specified, all Bagalayos quotes in this chapter are from this email correspondence.

6   *"it must be counted"*: René Descartes, *The Philosophical Writings of Descartes*, vol. 2 (1644; Cambridge: Cambridge University Press, 1984).

7   *in order to minimize surprise*: Karl Friston, "The Free-Energy Principle: A Unified Brain Theory?," *Nature Reviews Neuroscience* 11, no. 2 (February 2010): 127–38, https://doi.org/10.1038/nrn2787.

8   *"provide the owner organism"*: Antonio Damasio, *Feeling and Knowing: Making Minds Conscious* (New York: Knopf, 2021).

12 *slim report to Congress*: Department of Defense, *National Security Implications of Climate-Related Risks and a Changing Climate*, July 23, 2015.

13 *between climate and conflict*: Marshall Burke, Solomon M. Hsiang, and Edward Miguel, "Climate and Conflict," *Annual Review of Economics* 7, no. 1 (2015): 577–617, https://doi.org/10.1146/annurev-economics-080614 -115430.

15 *"a slow-moving issue"*: Julie Hirschfeld Davis, Mark Landler, and Coral Davenport, "Obama on Climate Change: The Trends Are 'Terrifying,'" *New York Times*, September 8, 2016, https://www.nytimes.com/2016/09/08/us/politics /obama-climate-change.html.

15 *has summers like Tampa's*: David Leonhardt, "Extreme Summer," *New York Times*, July 20, 2021, sec. Briefing, https://www.nytimes.com/2021/07/20 /briefing/heatwave-american-west-climate-change.html.

1. A HISTORY OF FORGETTING

21 *bones in the print*: A. Feldman, "A Sketch of the Technical History of Radiology from 1896 to 1920," *RadioGraphics* 9, no. 6 (November 1989): 1113–28, https://doi.org/10.1148/radiographics.9.6.2685937.

22 *"glass of a hothouse"*: Svante Arrhenius, "On the Influence of Carbonic Acid in the Air upon the Temperature of the Ground," *London, Edinburgh, and Dublin Philosophical Magazine and Journal of Science* 41, no. 251 (April 1, 1896): 237–76, https://doi.org/10.1080/14786449608620846.

22 *"be a rather stable thing"*: J. B. Kincer, "Is Our Climate Changing? A Study of Long-Time Temperature Trends," *Monthly Weather Review* 61, no. 9 (September 29, 1933): 251–59, https://doi.org/10.1175/1520-0493(1933)61<251:IOC CAS>2.0.CO;2.

23 *peek at the previous* century: J. B. Kincer, "Our Changing Climate," *Bulletin of the American Meteorological Society* 20, no. 10 (1939): 448–50.

24 *"an impending reversal"*: J. B. Kincer, "Our Changing Climate," *Eos, Transactions, American Geophysical Union* 27, no. 3 (1946): 342–47, https://doi.org/10 .1029/TR027i003p00342.

25 *"thousands or millions of years"*: World Meteorological Organization, "FAQs— Climate," 2022, https://public.wmo.int/en/about-us/frequently-asked -questions/climate.

26 *"unruly and unpredictable weather"*: Mike Hulme, *Weathered: Cultures of Climate* (London: SAGE Publications, 2017).

27 *"a metacognitive role"*: Trevor A. Harley, "Nice Weather for the Time of Year: The British Obsession with the Weather," in *Weather, Climate, Culture*, ed. Sarah Strauss and Ben Orlove (Milton Park, Abingdon, Oxon, UK: Routledge, 2003).

27 *"a very simple mental process"*: Clara M. Hitchcock, "The Psychology of Expectation," *Psychological Review: Monograph Supplements* 5, no. 3 (1903): i—78, https://doi.org/10.1037/h0093000.

29 *didn't contain enough experiments*: Dorothy G. Rogers, *Women Philosophers*, vol. 2, *Entering Academia in Nineteenth-Century America* (London: Bloomsbury Academic, 2021).

30 *flies continued to avoid*: Jacob A. Berry et al., "Dopamine Is Required for Learning and Forgetting in Drosophila," *Neuron* 74, no. 3 (May 10, 2012): 530–42, https://doi.org/10.1016/j.neuron.2012.04.007.

30 *they could stop rats*: Paola Virginia Migues et al., "Blocking Synaptic Removal of GluA2-Containing AMPA Receptors Prevents the Natural Forgetting of Long-Term Memories," *Journal of Neuroscience* 36, no. 12 (March 23, 2016): 3481–94, https://doi.org/10.1523/JNEUROSCI.3333-15.2016.

31 *"forgetting as an afterthought"*: Lauren Gravitz, "The Forgotten Part of Memory," *Nature* 571, no. 7766 (July 24, 2019): S12–14, https://doi.org/10.1038/d41586-019-02211-5.

32 *"almost intolerably exact"*: Jorge Luis Borges, "Funes, the Memorious," in *Ficciones* (New York: Grove Press, 1962), 112.

33 *people who have lived through earthquakes*: Laura Piccardi et al., "Continuous Environmental Changes May Enhance Topographic Memory Skills. Evidence from L'Aquila Earthquake–Exposed Survivors," *Frontiers in Human Neuroscience* 12 (2018), https://www.frontiersin.org/article/10.3389/fnhum.2018.00318.

33 *the rate of forgetting*: Tomás J. Ryan and Paul W. Frankland, "Forgetting as a Form of Adaptive Engram Cell Plasticity," *Nature Reviews Neuroscience* (January 13, 2022): 1–14, https://doi.org/10.1038/s41583-021-00548-3.

33 *"not all memories"*: Ryan and Frankland, "Forgetting as a Form of Adaptive Engram Cell Plasticity."

34 *"disorients our memories"*: Mike Hulme. "Climate Change and Memory," in *Memory in the Twenty-First Century: New Critical Perspectives from the Arts, Humanities, and Sciences*, ed. Sebastian Groes (London: Palgrave Macmillan UK, 2016), 159–62.

34 *almost half the year*: Jiamin Wang et al., "Changing Lengths of the Four Seasons by Global Warming," *Geophysical Research Letters* 48, no. 6 (2021): e2020G L091753, https://doi.org/10.1029/2020GL091753.

36 *"normals are calculated retrospectively"*: Anthony Arguez and Russell S. Vose, "The Definition of the Standard WMO Climate Normal: The Key to Deriving Alternative Climate Normals," *Bulletin of the American Meteorological Society* 92, no. 6 (June 1, 2011): 699–704, https://doi.org/10.1175/2010BAMS 2955.1.

36 *"a seer to predict"*: Michael A. Palecki, "New US Climate Normals Are Arriving Soon," *The Hill*, April 23, 2021, https://thehill.com/opinion/energy-environ ment/549919-new-us-climate-normals-are-arriving-soon/.

37 *"gradual shift of the baseline"*: Daniel Pauly, "Anecdotes and the Shifting Baseline Syndrome of Fisheries," *Trends in Ecology and Evolution* 10, no. 10 (October 1995): 430, https://doi.org/10.1016/S0169-5347(00)89171-5.

38 *average 70 percent decline*: World Wildlife Fund, "The 2022 Living Planet Report," accessed December 1, 2022, https://livingplanet.panda.org/en-US/.

38 *fail to correctly estimate*: S. K. Papworth et al., "Evidence for Shifting Baseline Syndrome in Conservation," *Conservation Letters* 2, no. 2 (2009): 93–100, https://doi.org/10.1111/j.1755-263X.2009.00049.x. For further reading, see: Masashi Soga and Kevin J. Gaston, "Shifting Baseline Syndrome: Causes, Consequences, and Implications," *Frontiers in Ecology and the Environment* 16, no. 4 (May 2018): 222–30, https://doi.org/10.1002/fee.1794.

39 *sixteen cubic miles of ice*: Snaevarr Guðmundsson, Helgi Björnsson, and Finnur Pálsson, "Changes of Breiðamerkurjökull Glacier, SE-Iceland, from Its Late Nineteenth Century Maximum to the Present," *Geografiska Annaler: Series A, Physical Geography* 99 (July 27, 2017): 1–15, https://doi.org/10.1080/04353676.2017.1355216.

40 *"resulting in transient forgetting"*: John Martin Sabandal, Jacob A. Berry, and Ronald L. Davis, "Dopamine-Based Mechanism for Transient Forgetting," *Nature* 591, no. 7850 (March 2021): 426–30, https://doi.org/10.1038/s41586-020-03154-y.

40 *failed to quash*: Sabandal, Berry, and Davis, "Dopamine-Based Mechanism for Transient Forgetting."

41 *"as societies fragment"*: Mike Hulme et al., "Unstable Climates: Exploring the Statistical and Social Constructions of 'Normal' Climate," *Geoforum* 40 (March 1, 2009): 197–206, https://doi.org/10.1016/j.geoforum.2008.09.010.

41 *forgetting isn't necessarily permanent*: Ryan and Frankland, "Forgetting as a Form of Adaptive Engram Cell Plasticity."

42 *"testing of pertinent hypotheses"*: Pauly, "Anecdotes and the Shifting Baseline Syndrome of Fisheries."

42 *notes in her essay*: Leslie Jamison, "The Empathy Exams," Culture, February 1, 2014, https://culture.org/the-empathy-exams/.

43 *"borders on an obsession"*: Harley, "Nice Weather for the Time of Year."

44 *focused on the use of parable*: David Glassberg, "Place, Memory, and Climate Change," *Public Historian* 36, no. 3 (August 1, 2014): 17–30, https://doi.org/10.1525/tph.2014.36.3.17.

44 *"reality of the present situation"*: Glassberg, "Place, Memory, and Climate Change."

44 *emphasis on the "sickness"*: Ryan Hediger, *Homesickness: Of Trauma and the Longing for Place in a Changing Environment* (Minneapolis: University of Minnesota Press, 2019).

45 *"pre-determined dystopian future"*: Glassberg, "Place, Memory, and Climate Change."

45 *deregistered the glacier*: Arnar Árnason and Sigurjón Baldur Hafsteinsson, "A

Funeral for a Glacier: Mourning the More-Than-Human on the Edges of Modernity," *Thanatos* 9 (2020): 26.

2. WET MACHINES

49 *dry bulb thermometer in the room*: Norman H. Mackworth, "Effects of Heat on Wireless Operators," *British Journal of Industrial Medicine* 3, no. 3 (July 1946): 143–58.

51 *"to do as a teacher"*: Sophie Date, telephone interview with the author, November 27, 2019. Unless otherwise specified, all Date quotes in this chapter are from this interview.

51 *referred to the oceans*: Anthony R. Wood and Joseph A. Gambardello, "Heat Indexes Hit 105; Ocean Is 'Bath Water' . . . but First Totally Dry Week since November," *Philadelphia Inquirer*, August 28, 2018, https://www.inquirer.com /philly/news/philadelphia-weather-schools-first-day-heat-wave-20180827 .html.

51 *called the conditions "inhumane"*: Kristen A. Graham, "Philly Schools to Close Early Wednesday; Officials Decry 'Dangerous' Conditions inside Schools," *Philadelphia Inquirer*, September 4, 2018, https://www.inquirer.com/philly/ed ucation/philly-schools-close-early-dangerous-heat-conditions-20180904.html.

53 *breed smaller chunks of cortex*: Talia Sanders et al., "Neurotoxic Effects and Biomarkers of Lead Exposure: A Review," *Reviews on Environmental Health* 24, no. 1 (2009): 15–45.

53 *its long-term memory will suffer*: Camilla Soravia et al., "The Impacts of Heat Stress on Animal Cognition: Implications for Adaptation to a Changing Climate," *Wiley Interdisciplinary Reviews: Climate Change* 12 (July 1, 2021), https:// doi.org/10.1002/wcc.713.

53 *"hard to tell your story"*: Joshua Graff Zivin, telephone interview with the author, November 14, 2019. Unless otherwise specified, all Graff Zivin quotes in this chapter are from this interview.

55 *productivity fell by more than 5 percent*: Joshua Graff Zivin and Matthew Neidell, "The Impact of Pollution on Worker Productivity," *American Economic Review* 102, no. 7 (December 1, 2012): 3652–73, https://doi.org/10.1257/aer .102.7.3652.

56 *pollution slowed down everyone*: Tom Chang et al., "Particulate Pollution and the Productivity of Pear Packers," *American Economic Journal: Economic Policy* 8, no. 3 (August 2016): 141–69, https://doi.org/10.1257/pol.20150085. For further reading, see: Tom Y. Chang et al., "The Effect of Pollution on Worker Productivity: Evidence from Call Center Workers in China," *American Eco-*

nomic Journal: Applied Economics 11, no. 1 (January 2019): 151–72, https://doi .org/10.1257/app.20160436.

56 *tanked students' math scores*: Joshua Graff Zivin, Solomon M. Hsiang, and Matthew Neidell, "Temperature and Human Capital in the Short and Long Run," *Journal of the Association of Environmental and Resource Economists* 5, no. 1 (January 2018): 77–105, https://doi.org/10.1086/694177.

56 *a heat wave in Boston*: Jose Guillermo Cedeño Laurent et al., "Reduced Cognitive Function during a Heat Wave among Residents of Non-air-conditioned Buildings: An Observational Study of Young Adults in the Summer of 2016," *PLOS Medicine* 15, no. 7 (July 10, 2018): e1002605, https://doi.org/10.1371 /journal.pmed.1002605.

57 *the cognitive effects of hotter days*: R. Jisung Park et al., "Heat and Learning," *American Economic Journal: Economic Policy* 12, no. 2 (May 2020): 306–39, https://doi.org/10.1257/pol.20180612.

57 *a 1 percent decrease in gaokao scores*: Joshua Graff Zivin et al., "Temperature and High-Stakes Cognitive Performance: Evidence from the National College Entrance Examination in China," *Journal of Environmental Economics and Management* 104 (November 1, 2020): 102365, https://doi.org/10.1016/j.jeem.2020 .102365.

58 *slumped with rising particulate-matter levels*: Avraham Ebenstein, Victor Lavy, and Sefi Roth, "The Long-Run Economic Consequences of High-Stakes Examinations: Evidence from Transitory Variation in Pollution," *American Economic Journal: Applied Economics* 8, no. 4 (October 2016): 36–65, https://doi .org/10.1257/app.20150213.

59 *"climate just affects everything"*: Solomon Hsiang, interview with the author, Berkeley, California, January 13, 2020. Unless otherwise specified, all Hsiang quotes in this chapter are from this interview.

59 *"contemporaneous temperature"*: Anthony Heyes and Soodeh Saberian, "Temperature and Decisions: Evidence from 207,000 Court Cases," *American Economic Journal: Applied Economics* 11, no. 2 (2019): 238–65, https://doi.org/10 .1257/app.20170223.

60 *"7.9 percent"*: Heyes and Saberian, "Temperature and Decisions."

60 *"subtle and pernicious"*: Heyes and Saberian, "Temperature and Decisions."

61 *swathes of evolutionarily new cortex*: Erica A. Boschin et al., "Distinct Roles for the Anterior Cingulate and Dorsolateral Prefrontal Cortices during Conflict between Abstract Rules," *Cerebral Cortex* 27, no. 1 (January 1, 2017): 34–45, https://doi.org/10.1093/cercor/bhw350.

63 *press the appropriate arrow key*: Nadia Gaoua et al., "Effect of Passive Hyperthermia on Working Memory Resources during Simple and Complex Cogni-

tive Tasks," *Frontiers in Psychology* 8 (January 11, 2018): 2290, https://doi.org /10.3389/fpsyg.2017.02290.

63 *exposed volunteers to extreme heat*: Shaowen Qian et al., "Effects of Short-Term Environmental Hyperthermia on Patterns of Cerebral Blood Flow," *Physiology and Behavior* 128 (April 2014): 99–107, https://doi.org/10.1016 /j.physbeh.2014.01.028.

63 *heat reduces the connectivity*: Gang Sun et al., "Hyperthermia-Induced Disruption of Functional Connectivity in the Human Brain Network," *PLOS ONE* 8, no. 4 (April 8, 2013): e61157, https://doi.org/10.1371/journal.pone.0061157. For further reading, see: Shaowen Qian et al., "Altered Topological Patterns of Large-Scale Brain Functional Networks during Passive Hyperthermia," *Brain and Cognition* 83, no. 1 (October 2013): 121–31, https://doi.org/10.1016 /j.bandc.2013.07.013.

64 *the average July temperature*: Eugene A. Kiyatkin and Hari S. Sharma, "Expression of Heat Shock Protein (HSP 72 KD) during Acute Methamphetamine Intoxication Depends on Brain Hyperthermia: Neurotoxicity or Neuroprotection?," *Journal of Neural Transmission* 118, no. 1 (January 2011): 47–60, https:// doi.org/10.1007/s00702-010-0477-5. For further reading, see: Eugene A. Kiyatkin, "Brain Temperature Homeostasis: Physiological Fluctuations and Pathological Shifts," *Frontiers in Bioscience: A Journal and Virtual Library* 15 (January 1, 2010): 73–92.

65 *called an alliesthesial effect*: Michel Cabanac, "Physiological Role of Pleasure: A Stimulus Can Feel Pleasant or Unpleasant Depending upon Its Usefulness as Determined by Internal Signals," *Science* 173, no. 4002 (1971): 1103–7.

67 *"environment in which they function"*: Heyes and Saberian, "Temperature and Decisions."

67 *oft-quoted 1999 interview*: Lee Kuan Yew, "The East Asian Way—with Air Conditioning," *New Perspectives Quarterly* 26, no. 4 (September 2009): 111–20, https://doi.org/10.1111/j.1540-5842.2009.01120.x.

68 *strapping cold packs to people's foreheads*: S. Racinais, N. Gaoua, and J. Grantham, "Hyperthermia Impairs Short-Term Memory and Peripheral Motor Drive Transmission," *Journal of Physiology* 586, no. 19 (2008): 4751–62, https://doi.org/10.1113/jphysiol.2008.157420.

68 *looking at images of cool environments*: Eliran Halali, Nachshon Meiran, and Idit Shalev, "Keep It Cool: Temperature Priming Effect on Cognitive Control," *Psychological Research* 81, no. 2 (March 1, 2017): 343–54, https://doi.org /10.1007/s00426-016-0753-6.

68 *a third of the global population*: Chi Xu et al., "Future of the Human Cli-

mate Niche," *Proceedings of the National Academy of Sciences* 117, no. 21 (2020): 11350–55.

70 *"something you can't ignore here"*: Christiana Moss, telephone interview with the author, September 16, 2021. Unless otherwise specified, all Moss quotes in this chapter are from this interview.

71 *windcatchers in the Middle East*: David Hambling, "Ancient Windcatchers in Iran Give Architects Cooling Inspiration," *Guardian*, July 13, 2023, https:// www.theguardian.com/environment/2023/jul/13/weatherwatch-ancient -windcatchers-iran-give-architects-cooling-inspiration.

3. WHO KILLED TYSON MORLOCK?

74 *Morlock had told a journalist*: Savannah Eadens, "Tyson Morlock, Fatally Stabbed by Friend, One of Several Killed in 2021 While Living on Portland Streets," OregonLive, October 9, 2021, https://www.oregonlive.com/data /2021/10/tyson-morlock-fatally-stabbed-by-friend-one-of-several-killed-in -2021-while-living-on-portland-streets.html.

76 *collecting wild* P. moluccensis *from coral reefs*: Peter A. Biro, Christa Beck-mann, and Judy A. Stamps, "Small Within-Day Increases in Temperature Af-fects Boldness and Alters Personality in Coral Reef Fish," *Proceedings of the Royal Society B: Biological Sciences* 277, no. 1678 (September 30, 2009): 71–77, https://doi.org/10.1098/rspb.2009.1346.

78 *inclined toward sibling cannibalism*: Christopher de Tranaltes et al., "Sibli-cide in the City: The Urban Heat Island Accelerates Sibling Cannibalism in the Black Widow Spider (*Latrodectus hesperus*)," *Urban Ecosystems* 25, no. 1 (February 1, 2022): 305–12, https://doi.org/10.1007/s11252-021-01148-w.

78 *seize and drag workers*: Patrick Krapf et al., "Global Change May Make Hostile—Higher Ambient Temperature and Nitrogen Availability Increase Ant Aggression," *Science of the Total Environment* 861 (February 25, 2023): 160443, https://doi.org/10.1016/j.scitotenv.2022.160443.

78 *the monkeys start more fights*: Aichun Xu et al., "Monkeys Fight More in Pol-luted Air," *Scientific Reports* 11, no. 1 (January 12, 2021): 654, https://doi.org /10.1038/s41598-020-80002-z.

78 *every pitch, every outcome*: Richard P. Larrick et al., "Temper, Temperature, and Temptation: Heat-Related Retaliation in Baseball," *Psychological Science* 22, no. 4 (April 2011): 423–28, https://doi.org/10.1177/0956797611399292.

80 *begins with an axe murder*: Fyodor Dostoyevsky, *Crime and Punishment* (1866; Oxford: Oxford University Press, 2017).

80 *wrote of scorching heat*: Albert Camus, *The Stranger* (1946; New York: Vintage, 2012).

80 *the retired insurance salesman*: Ray Bradbury, "Touched with Fire," in *The October Country* (New York: Ballantine, 1955).

81 *a scorching red sun*: 王維 (Wang Wei), "苦熱行" ("Ballad of Suffering from the Heat"), trans. East Asia Student, accessed August 1, 2023, https://eastasia student.net/china/classical/wang-wei-ku-re-xing/.

81 *rife with historical evidence*: J. Merrill Carlsmith and Craig A. Anderson, "Ambient Temperature and the Occurrence of Collective Violence: A New Analysis," *Journal of Personality and Social Psychology* 37 (1979): 337–44, https://doi.org/10.1037/0022-3514.37.3.337. For further reading, see: Alexander Henke and Lin-chi Hsu, "The Gender Wage Gap, Weather, and Intimate Partner Violence," *Review of Economics of the Household* 18, no. 2 (June 1, 2020): 413–29, https://doi.org/10.1007/s11150-020-09483-1. Also: Keith D. Harries and Stephen J. Stadler, "Determinism Revisited: Assault and Heat Stress in Dallas, 1980," *Environment and Behavior* 15, no. 2 (March 1, 1983): 235–56, https://doi.org/10.1177/0013916583152006.

81 *"play in exacerbating conflict"*: Larrick et al., "Temper, Temperature, and Temptation."

82 *a precise and exacting study*: Ayushi Narayan, "The Impact of Extreme Heat on Workplace Harassment and Discrimination," *Proceedings of the National Academy of Sciences* 119, no. 39 (September 27, 2022): e2204076119, https://doi.org/10.1073/pnas.2204076119.

83 *"may not be enough"*: Ayushi Narayan, telephone interview with the author, February 2, 2023. Unless otherwise specified, all Narayan quotes in this chapter are from this interview.

83 *likelihood of civil riots*: Carlsmith and Anderson, "Ambient Temperature and the Occurrence of Collective Violence."

84 *"replicate this in a lab?"*: Craig Anderson, telephone interview with the author, October 27, 2015. Unless otherwise specified, all Anderson quotes in this chapter are from this interview.

84 *fascinated with Westerns*: Craig A. Anderson, "Human Aggression and Violence," in *Scientists Making a Difference: One Hundred Eminent Behavioral and Brain Scientists Talk about Their Most Important Contributions*, ed. Robert J. Sternberg, Susan T. Fiske, and Donald J. Foss (New York: Cambridge University Press, 2016).

86 *turned their attention to drivers*: Douglas T. Kenrick and Steven W. MacFarlane, "Ambient Temperature and Horn Honking: A Field Study of the Heat/Aggression Relationship," *Environment and Behavior* 18, no. 2 (1986): 179–91, https://doi.org/10.1177/0013916586182002.

88 *sought to rectify this problem*: Jari Tiihonen et al., "The Association of

Ambient Temperature and Violent Crime," *Scientific Reports* 7, no. 1 (July 28, 2017): 6543, https://doi.org/10.1038/s41598-017-06720-z.

89 *"Cold weather is the best police"*: Jari Tiihonen, email correspondence with the author, December 28, 2022.

91 *repeat the absolute numbers*: Matthew Ranson, "Crime, Weather, and Climate Change," *Journal of Environmental Economics and Management* 67, no. 3 (2014): 274–302.

92 *on the order of 350 milliseconds*: Benjamin Libet et al., "Time of Conscious Intention to Act in Relation to Onset of Cerebral Activity (Readiness-Potential): The Unconscious Initiation of a Freely Voluntary Act," *Brain* 106, no. 3 (September 1, 1983): 623–42, https://doi.org/10.1093/brain/106.3.623. For further reading, see: Moritz Nicolai Braun, Janet Wessler, and Malte Friese, "A Meta-analysis of Libet-Style Experiments," *Neuroscience and Biobehavioral Reviews* 128 (September 1, 2021): 182–98, https://doi.org/10.1016/j.neubiorev.2021.06.018.

95 *brain serotonin transmission and impulsivity*: B. U. Phillips and T. W. Robbins, "The Role of Central Serotonin in Impulsivity, Compulsivity, and Decision-Making: Comparative Studies in Experimental Animals and Humans," in *Handbook of Behavioral Neuroscience*, ed. Christian P. Müller and Kathryn A. Cunningham (London: Elsevier, 2020), 31:531–48, https://doi.org/10.1016/B978-0-444-64125-0.00031-1.

96 *explain impulsive aggressive behavior*: Sofi da Cunha-Bang and Gitte Moos Knudsen, "The Modulatory Role of Serotonin on Human Impulsive Aggression," *Biological Psychiatry* 90, no. 7 (October 2021): 447–57, https://doi.org/10.1016/j.biopsych.2021.05.016.

97 *without having cortical inhibition*: Brent A. Vogt, "Cingulate Impairments in ADHD: Comorbidities, Connections, and Treatment," in *Handbook of Clinical Neurology* 116: 297–314, https://doi.org/10.1016/B978-0-444-64196-0.00016-9.

97 *ADHD medications act differentially*: Ciarán M. Fitzpatrick and Jesper T. Andreasen, "Differential Effects of ADHD Medications on Impulsive Action in the Mouse 5-Choice Serial Reaction Time Task," *European Journal of Pharmacology* 847 (March 15, 2019): 123–29, https://doi.org/10.1016/j.ejphar.2019.01.038.

97 *tend to produce impulsive responses*: Miranda L. Virone, "The Use of Mindfulness to Improve Emotional Regulation and Impulse Control among Adolescents with ADHD," *Journal of Occupational Therapy, Schools, and Early Intervention* 16, no. 1 (January 2, 2023): 78–90, https://doi.org/10.1080/19411243.2021.2009081.

4. BLOOM

102 *had been found marooned*: David A. Davis et al., "Cyanobacterial Neurotoxin BMAA and Brain Pathology in Stranded Dolphins," *PLOS ONE* 14, no. 3 (March 20, 2019): e0213346, https://doi.org/10.1371/journal.pone.0213346.

104 *"look like they have Alzheimer's disease"*: David Davis, telephone interview with the author, August 26, 2022. Unless otherwise specified, all Davis quotes in this chapter are from this interview.

105 *converting light into chemical energy*: Tanai Cardona et al., "Early Archean Origin of Photosystem II," *Geobiology* 17, no. 2 (March 2019): 127–50, https://doi.org/10.1111/gbi.12322.

106 *nab protons from the hydrogen*: Wolfgang Lubitz, Maria Chrysina, and Nicholas Cox, "Water Oxidation in Photosystem II," *Photosynthesis Research* 142, no. 1 (October 1, 2019): 105–25, https://doi.org/10.1007/s11120-019-00648-3.

106 *Earth cooled to a snowball*: Robert E. Kopp et al., "The Paleoproterozoic Snowball Earth: A Climate Disaster Triggered by the Evolution of Oxygenic Photosynthesis," *Proceedings of the National Academy of Sciences* 102, no. 32 (August 9, 2005): 11131–36, https://doi.org/10.1073/pnas.0504878102. For further reading, see: Matthew R. Warke et al., "The Great Oxidation Event Preceded a Paleoproterozoic 'Snowball Earth,'" *Proceedings of the National Academy of Sciences* 117, no. 24 (June 16, 2020): 13314–20, https://doi.org/10.1073/pnas.2003090117.

106 *as cold as Antarctica*: William T. Hyde et al., "Neoproterozoic 'Snowball Earth' Simulations with a Coupled Climate/Ice-Sheet Model," *Nature* 405, no. 6785 (May 2000): 425–29, https://doi.org/10.1038/35013005.

106 *tend to lie dormant*: Kathryn L. Cottingham et al., "Predicting the Effects of Climate Change on Freshwater Cyanobacterial Blooms Requires Consideration of the Complete Cyanobacterial Life Cycle," *Journal of Plankton Research* 43, no. 1 (January 1, 2021): 10–19, https://doi.org/10.1093/plankt/fbaa059.

107 *half the size of Germany*: M. Kahru and R. Elmgren, "Multidecadal Time Series of Satellite-Detected Accumulations of Cyanobacteria in the Baltic Sea," *Biogeosciences* 11, no. 13 (July 4, 2014): 3619–33, https://doi.org/10.5194/bg -11-3619-2014.

108 *contained a compound entirely new*: Wendee Holtcamp, "The Emerging Science of BMAA: Do Cyanobacteria Contribute to Neurodegenerative Disease?" *Environmental Health Perspectives* 120, no. 3 (March 2012): a110–16, https://doi.org/10.1289/ehp.120-a110. For further reading, see: A. Vega and E. A. Bell, "α-Amino-β-Methylaminopropionic Acid, a New Amino Acid from Seeds of Cycas Circinalis," *Phytochemistry* 6, no. 5 (May 1, 1967): 759–62, https://doi.org/10.1016/S0031-9422(00)86018-5.

108 *chronic exposure that mattered*: Paul Alan Cox and Oliver W. Sacks, "Cycad Neurotoxins, Consumption of Flying Foxes, and ALS-PDC Disease in Guam," *Neurology* 58, no. 6 (March 26, 2002): 956–59, https://doi.org/10.1212/WNL .58.6.956.

109 *detecting enormous levels of BMAA*: Paul Alan Cox, Sandra Anne Banack, and Susan J. Murch, "Biomagnification of Cyanobacterial Neurotoxins and Neu-rodegenerative Disease among the Chamorro People of Guam," *Proceedings of the National Academy of Sciences* 100, no. 23 (November 11, 2003): 13380–83, https://doi.org/10.1073/pnas.2235808100. For further reading, see: Sandra Anne Banack and Paul Alan Cox, "Biomagnification of Cycad Neurotoxins in Flying Foxes: Implications for ALS-PDC in Guam," *Neurology* 61, no. 3 (August 12, 2003): 387–89, https://doi.org/10.1212/01.WNL.0000078320.185 64.9F.

109 *shark fins and shark muscles*: Larry E. Brand et al., "Cyanobacterial Blooms and the Occurrence of the Neurotoxin, Beta-N-Methylamino-l-Alanine (BMAA), in South Florida Aquatic Food Webs," *Harmful Algae* 9, no. 6 (September 1, 2010): 620–35, https://doi.org/10.1016/j.hal.2010.05.002. For further reading, see: Kiyo Mondo et al., "Cyanobacterial Neurotoxin β-N -Methylamino-L-Alanine (BMAA) in Shark Fins," *Marine Drugs* 10, no. 2 (February 2012): 509–20, https://doi.org/10.3390/md10020509.

110 *researchers at the brain bank*: J. Pablo et al., "Cyanobacterial Neurotoxin BMAA in ALS and Alzheimer's Disease," *Acta Neurologica Scandinavica* 120, no. 4 (2009): 216–25, https://doi.org/10.1111/j.1600-0404.2008.01150.x.

110 *dolphins the researchers examined*: Davis et al., "Cyanobacterial Neurotoxin BMAA and Brain Pathology in Stranded Dolphins."

111 *cyanobacteria outpacing the growth*: Zofia E. Taranu et al., "Acceleration of Cyanobacterial Dominance in North Temperate-Subarctic Lakes during the Anthropocene," *Ecology Letters* 18, no. 4 (2015): 375–84.

111 *witnessed increased bloom intensity*: Jeff C. Ho, Anna M. Michalak, and Nima Pahlevan, "Widespread Global Increase in Intense Lake Phytoplankton Blooms since the 1980s," *Nature* 574, no. 7780 (October 2019): 667–70, https://doi.org /10.1038/s41586-019-1648-7.

111 *offer a nutrient buffet*: Jennifer Jankowiak et al., "Deciphering the Effects of Nitrogen, Phosphorus, and Temperature on Cyanobacterial Bloom Intensification, Diversity, and Toxicity in Western Lake Erie," *Limnology and Oceanography* 64, no. 3 (2019): 1347–70, https://doi.org/10.1002/lno.11120. For further reading, see: Andrew M. Dolman et al., "Cyanobacteria and Cyanotoxins: The Influence of Nitrogen versus Phosphorus," *PLOS ONE* 7, no. 6 (June 15, 2012): e38757, https://doi.org/10.1371/journal.pone.0038757.

111 *goes unused by the crops*: Paul C. West et al., "Leverage Points for Improving Global Food Security and the Environment," *Science* 345, no. 6194 (July 18, 2014): 325–28, https://doi.org/10.1126/science.1246067.

112 *carbon dioxide–rich waters*: Hadayet Ullah et al., "Climate Change Could Drive Marine Food Web Collapse through Altered Trophic Flows and Cyano- bacterial Proliferation," *PLOS Biology* 16, no. 1 (January 9, 2018): e2003446, https://doi.org/10.1371/journal.pbio.2003446.

114 *clinical trials of drugs*: Jeffrey Cummings, "Lessons Learned from Alzheimer Disease: Clinical Trials with Negative Outcomes," *Clinical and Translational Science* 11, no. 2 (March 2018): 147–52, https://doi.org/10.1111/cts.12491.

114 *"takes everything away from you"*: Elijah Stommel, "Risk Factors for Amyo- trophic Lateral Sclerosis: Cyanobacteria and Others," lecture to NH Health- care Workers for Climate Action, July 13, 2022, https://www.youtube.com /watch?v=SFYqcRZMabU.

115 *cases surrounding the lake*: Tracie A. Caller et al., "A Cluster of Amyotrophic Lateral Sclerosis in New Hampshire: A Possible Role for Toxic Cyanobacteria Blooms," *Amyotrophic Lateral Sclerosis* 10, no. S2 (January 1, 2009): 101–8, https://doi.org/10.3109/17482960903278485.

116 *"two to three minutes"*: Elijah Stommel, telephone interview with the author, August 14, 2022. Unless otherwise specified, all Stommel quotes in this chap- ter are from this interview.

116 *blooms cause locomotor deficits*: Jiaming Hu et al., "Exposure to Aerosolized Algal Toxins in South Florida Increases Short- and Long-Term Health Risk in Drosophila Model of Aging," *Toxins* 12, no. 12 (December 2020): 787, https:// doi.org/10.3390/toxins12120787.

116 *BMAA targets the olfactory bulb*: Paula Pierozan, Daiane Cattani, and Oskar Karlsson, "Hippocampal Neural Stem Cells Are More Susceptible to the Neu- rotoxin BMAA than Primary Neurons: Effects on Apoptosis, Cellular Differ- entiation, Neurite Outgrowth, and DNA Methylation," *Cell Death and Disease* 11, no. 10 (2020): 910.

117 *detected BMAA in carp*: Sandra Anne Banack et al., "Detection of Cyanotox- ins, β-N-Methylamino-L-Alanine and Microcystins, from a Lake Surrounded by Cases of Amyotrophic Lateral Sclerosis," *Toxins* 7, no. 2 (February 2015): 322–36, https://doi.org/10.3390/toxins7020322.

118 *crusts and mats contained BMAA*: Paul Alan Cox et al., "Cyanobacteria and BMAA Exposure from Desert Dust: A Possible Link to Sporadic ALS among Gulf War Veterans," *Amyotrophic Lateral Sclerosis* 10, no. S2 (2009): 109–17.

118 *ALS in the French Alps*: E. Lagrange et al., "An Amyotrophic Lateral Sclerosis Hot Spot in the French Alps Associated with Genotoxic Fungi," *Journal of the*

Neurological Sciences 427 (August 15, 2021): 117558, https://doi.org/10.1016 /j.jns.2021.117558.

118 *broadcasted the title of a 2018 paper*: Niam M. Abeysiriwardena, Samuel J. L. Gascoigne, and Angela Anandappa, "Algal Bloom Expansion Increases Cyanotoxin Risk in Food," *Yale Journal of Biology and Medicine* 91, no. 2 (June 2018): 129–42.

118 *218 million people in the country*: US Environmental Protection Agency, *Atlas of America's Polluted Waters* (Washington, DC: 2000).

118 *"It does scare me"*: James Metcalf, interview with the author, Jackson, Wyoming, September 16, 2022.

118 *"astounding to me"*: John Cassani, telephone interview with the author, July 27, 2022.

119 *ingesting more methylmercury*: Amina T. Schartup et al., "Climate Change and Overfishing Increase Neurotoxicant in Marine Predators," *Nature* 572, no. 7771 (August 2019): 648–50, https://doi.org/10.1038/s41586-019-1468-9.

120 *paralysis-causing ocean toxins*: Christina C. Roggatz et al., "Saxitoxin and Tetrodotoxin Bioavailability Increases in Future Oceans," *Nature Climate Change* 9, no. 11 (November 2019): 840–44, https://doi.org/10.1038/s41558-019-0589-3.

120 *my Grist colleague Zoya Teirstein*: Zoya Teirstein, "Alaskan Roulette: As Warming Waters Make Shellfish Toxic, a Way of Life Becomes Deadly for Native Alaskans," Grist, February 25, 2020, https://grist.org/food/climate -change-is-turning-shellfish-toxic-and-threatening-alaska-natives/.

121 *"mediate species interactions"*: Roggatz et al., "Saxitoxin and Tetrodotoxin Bioavailability Increases in Future Oceans."

121 *stripping the insulating sheaths*: Jennifer M. Panlilio, Neelakanteswar Aluru, and Mark E. Hahn, "Developmental Neurotoxicity of the Harmful Algal Bloom Toxin Domoic Acid: Cellular and Molecular Mechanisms Underlying Altered Behavior in the Zebrafish Model," *Environmental Health Perspectives* 128, no. 11 (2020): 117002, https://doi.org/10.1289/EHP6652.

122 *releasing mercury into the water*: Jon R. Hawkings et al., "Large Subglacial Source of Mercury from the Southwestern Margin of the Greenland Ice Sheet," *Nature Geoscience* 14, no. 7 (July 2021): 496–502, https://doi.org/10 .1038/s41561-021-00753-w.

122 *cartoonish sludge zombie*: Kevin Schaefer et al., "Potential Impacts of Mercury Released from Thawing Permafrost," *Nature Communications* 11, no. 1 (December 2020): 4650, https://doi.org/10.1038/s41467-020-18398-5.

122 *density of tangled proteins*: David A. Davis et al., "BMAA, Methylmercury, and Mechanisms of Neurodegeneration in Dolphins: A Natural Model of

Toxin Exposure," *Toxins* 13, no. 10 (October 2021): 697, https://doi.org/10.3390/toxins13100697.

123 *statistical relationship must arise*: P. Armitage and R. Doll, "The Age Distribution of Cancer and a Multi-stage Theory of Carcinogenesis," *British Journal of Cancer* 8, no. 1 (March 1954): 1–12. For further reading, see: Chris Hornsby, Karen M. Page, and Ian P. M. Tomlinson, "What Can We Learn from the Population Incidence of Cancer? Armitage and Doll Revisited," *Lancet Oncology* 8, no. 11 (November 1, 2007): 1030–38, https://doi.org/10.1016/S1470-2045(07)70343-1.

124 *causally explain geographic clusters*: Elisa Longinetti et al., "Geographical Clusters of Amyotrophic Lateral Sclerosis and the Bradford Hill Criteria," *Amyotrophic Lateral Sclerosis and Frontotemporal Degeneration* 23, nos. 5–6 (2022): 329–43.

125 *the matador Julio Aparicio*: Juan Moreno, "Gored through the Neck: Matador Who Cheated Death Makes His Comeback," *Der Spiegel*, August 13, 2010, sec. International, https://www.spiegel.de/international/zeitgeist/gored-through-the-neck-matador-who-cheated-death-makes-his-comeback-a-711092.html.

128 *"eat fish or not"*: Manuel Aparicio, interview with the author, Fort Myers, Florida, August 23, 2022.

5. SPILLING

131 *tells the story of Philip's death*: Erin Welsh and Erin Allmann Updyke, "Episode 74 Naegleria fowleri: The 'Brain-Eating Amoeba,'" June 1, 2021, *This Podcast Will Kill You*, podcast, https://thispodcastwillkillyou.com/2021/06/01/episode-74-naegleria-fowleri-the-brain-eating-amoeba/.

133 *sports little intertube-like appendages*: Hae-Jin Sohn et al., "The Nf-Actin Gene Is an Important Factor for Food-Cup Formation and Cytotoxicity of Pathogenic *Naegleria fowleri*," *Parasitology Research* 106, no. 4 (March 1, 2010): 917–24, https://doi.org/10.1007/s00436-010-1760-y.

135 *contracted at a natural hot spring*: Rebekah Riess, "2-Year-Old Nevada Boy Dies from Brain-Eating Amoeba Likely Contracted at Natural Hot Spring," CNN, July 22, 2023, https://www.cnn.com/2023/07/22/us/nevada-brain-eating-amoeba/index.html.

135 *break down your nasal mucus*: Eddie Grace, Scott Asbill, and Kris Virga, "*Naegleria fowleri*: Pathogenesis, Diagnosis, and Treatment Options," *Antimicrobial Agents and Chemotherapy* 59, no. 11 (November 2015): 6677–81, https://doi.org/10.1128/AAC.01293-15.

137 *revealed to have animal roots:* Paul M. Sharp and Beatrice H. Hahn, "Origins of HIV and the AIDS Pandemic," *Cold Spring Harbor Perspectives in Medicine* 1, no. 1 (September 2011): a006841, https://doi.org/10.1101/cshperspect.a006841.

137 *responsible for nearly 3 million:* Md. Tanvir Rahman et al., "Zoonotic Diseases: Etiology, Impact, and Control," *Microorganisms* 8, no. 9 (September 12, 2020): 1405, https://doi.org/10.3390/microorganisms8091405.

138 *selectivity of the blood-brain barrier:* N. Joan Abbott et al., "Structure and Function of the Blood–Brain Barrier," *Neurobiology of Disease* 37, no. 1 (2010): 13–25.

139 *There is Japanese encephalitis:* Usha Kant Misra and Jayantee Kalita, "Overview: Japanese Encephalitis," *Progress in Neurobiology* 91, no. 2 (2010): 108–20.

139 *range of the* Culex: Valerie Hongoh et al., "Expanding Geographical Distribution of the Mosquito, *Culex pipiens*, in Canada under Climate Change," *Applied Geography* 33 (2012): 53–62.

139 *yellow fever:* Thomas P. Monath and Pedro F. C. Vasconcelos, "Yellow Fever," *Journal of Clinical Virology* 64 (2015): 160–73.

140 *known as the yellow fever mosquito:* Ana Cláudia Piovezan-Borges et al., "Global Trends in Research on the Effects of Climate Change on *Aedes aegypti*: International Collaboration Has Increased, but Some Critical Countries Lag Behind," *Parasites and Vectors* 15, no. 1 (2022): 1–12.

140 *There is Zika:* Didier Musso and Duane J. Gubler, "Zika Virus," *Clinical Microbiology Reviews* 29, no. 3 (2016): 487–524.

140 *which flourish in pockets:* Piovezan-Borges et al., "Global Trends in Research on the Effects of Climate Change on *Aedes aegypti*."

140 *There is cerebral malaria:* Richard Idro, Neil E. Jenkins, and Charles R. J. C. Newton, "Pathogenesis, Clinical Features, and Neurological Outcome of Cerebral Malaria," *Lancet Neurology* 4, no. 12 (2005): 827–40.

141 *carried by* Anopheles *mosquitoes:* Sadie J. Ryan et al., "Mapping Current and Future Thermal Limits to Suitability for Malaria Transmission by the Invasive Mosquito *Anopheles stephensi*," *Malaria Journal* 22, no. 1 (2023): 104.

141 *Powassan virus:* Meghan E. Hermance and Saravanan Thangamani, "Powassan Virus: An Emerging Arbovirus of Public Health Concern in North America," *Vector-Borne and Zoonotic Diseases* 17, no. 7 (2017): 453–62.

141 *Powassan range is broadening:* Catherine Bouchard et al., "Increased Risk of Tick-Borne Diseases with Climate and Environmental Changes," *Canada Communicable Disease Report* 45, no. 4 (2019): 83.

141 *"these northern-tier states":* Christopher Rice, telephone interview with the author, February 7, 2023. Unless otherwise specified, all Rice quotes in this chapter are from this interview.

141 *practicing nasal ablution*: Muḥammad Jahangeer at al., *"Naegleria fowleri*: Sources of Infection, Pathophysiology, Diagnosis, and Management: A Review," *Clinical and Experimental Pharmacology and Physiology* 47, no. 2 (2020): 199–212.

142 *medical professionals don't suspect*: Lynne Eger and Morgan A. Pence, "The Brief Case: A Case of Primary Amebic Meningoencephalitis (PAM) after Exposure at a Splash Pad," *Journal of Clinical Microbiology* 61, no. 7 (July 20, 2023): e0126922, https://doi.org/10.1128/jcm.01269-22.

142 *cocktail of antifungal and antibiotic*: Andrea Güémez and Elisa García, "Primary Amoebic Meningoencephalitis by *Naegleria fowleri*: Pathogenesis and Treatments," *Biomolecules* 11, no. 9 (2021): 1320.

144 *"economic losses worldwide"*: Stephanie J. Salyer et al., "Prioritizing Zoonoses for Global Health Capacity Building—Themes from One Health Zoonotic Disease Workshops in 7 Countries, 2014–2016," *Emerging Infectious Diseases* 23, no. S1 (December 2017): S55–S64, https://doi.org/10.3201/eid2313.170418.

144 *virus gains entry*: Matthias J. Schnell et al., "The Cell Biology of Rabies Virus: Using Stealth to Reach the Brain," *Nature Reviews Microbiology* 8, no. 1 (2010): 51–61.

144 *"blood with their tongue"*: US Geological Survey, "Do Vampire Bats Really Exist?," USGS North American Bat Monitoring Program (NABat), 2022, https://www.usgs.gov/faqs/do-vampire-bats-really-exist.

145 *roosts are social networks*: Gerald S. Wilkinson, "The Social Organization of the Common Vampire Bat: I. Pattern and Cause of Association," *Behavioral Ecology and Sociobiology* 17 (1985): 111–21.

146 *rabies virus rides along*: Mark A. Hayes and Antoinette J. Piaggio, "Assessing the Potential Impacts of a Changing Climate on the Distribution of a Rabies Virus Vector," *PLOS ONE* 13, no. 2 (2018): e0192887.

147 *"one of the greatest failures"*: Victoria Pilkington, Sarai Mirjam Keestra, and Andrew Hill, "Global COVID-19 Vaccine Inequity: Failures in the First Year of Distribution and Potential Solutions for the Future," *Frontiers in Public Health* 10 (2022), https://www.frontiersin.org/articles/10.3389/fpubh.2022.821117.

148 *"portion of the phenotypic variation"*: Hanxin Zhang, Atif Khan, and Andrey Rzhetsky, "Gene-Environment Interactions Explain a Substantial Portion of Variability of Common Neuropsychiatric Disorders," *Cell Reports Medicine* 3, no. 9 (September 20, 2022), https://doi.org/10.1016/j.xcrm.2022.100736.

148 *common respiratory infections*: Jason Arunn Murugesu, "Covid-19 Linked to Higher Risk of Brain Conditions up to Two Years On," *New Scientist*, accessed May 4, 2023, https://www.newscientist.com/article/2334325-covid-19-linked-to-higher-risk-of-brain-conditions-up-to-two-years-on/.

148 *within six months*: Maxime Taquet et al., "6-Month Neurological and Psychiatric Outcomes in 236 379 Survivors of COVID-19: A Retrospective Cohort Study Using Electronic Health Records," *Lancet Psychiatry* 8, no. 5 (May 1, 2021): 416–27, https://doi.org/10.1016/S2215-0366(21)00084-5.

149 *risk factors associated*: Shreya Louis et al., "Impacts of Climate Change and Air Pollution on Neurologic Health, Disease, and Practice: A Scoping Review," *Neurology* 100, no. 10 (March 7, 2023): 474–83, https://doi.org/10.1212/WNL.0000000000201630.

149 *least pleasant study*: Lilian Calderón-Garcidueñas et al., "Hallmarks of Alzheimer Disease Are Evolving Relentlessly in Metropolitan Mexico City Infants, Children and Young Adults. APOE4 Carriers Have Higher Suicide Risk and Higher Odds of Reaching NFT Stage V at ≤ 40 Years of Age," *Environmental Research* 164 (July 1, 2018): 475–87, https://doi.org/10.1016/j.envres.2018.03.023.

149 *a largely heritable disorder*: E. Fuller Torrey and Robert H. Yolken, "Schizophrenia as a Pseudogenetic Disease: A Call for More Gene-Environmental Studies," *Psychiatry Research* 278 (2019): 146–50.

150 *shown to explain*: Lotta-Katrin Pries et al., "Estimating Aggregate Environmental Risk Score in Psychiatry: The Exposome Score for Schizophrenia," *Frontiers in Psychiatry* 12 (2021), https://www.frontiersin.org/articles/10.3389/fpsyt.2021.671334.

150 *model of heritability explained*: Hanxin Zhang et al., "Dissecting Schizophrenia Phenotypic Variation: The Contribution of Genetic Variation, Environmental Exposures, and Gene-Environment Interactions," *Schizophrenia* 8, no. 1 (May 10, 2022): 51, https://doi.org/10.1038/s41537-022-00257-5.

150 *developing neuropsychiatric conditions*: Jennifer Puthota et al., "Prenatal Ambient Temperature and Risk for Schizophrenia," *Schizophrenia Research* (October 5, 2021), https://doi.org/10.1016/j.schres.2021.09.020.

150 *climatic factors and neuropsychiatric health*: American Academy of Neurology, "Are Climate Change and Air Pollution Making Neurologic Diseases Worse?," *ScienceDaily*, November 16, 2022, www.sciencedaily.com/releases/2022/11/221116164916.htm.

150 *conditions of food insecurity*: Clara Y. Park and Heather A. Eicher-Miller, "Iron Deficiency Is Associated with Food Insecurity in Pregnant Females in the United States: National Health and Nutrition Examination Survey 1999–2010," *Journal of the Academy of Nutrition and Dietetics* 114, no. 12 (December 2014): 1967–73, https://doi.org/10.1016/j.jand.2014.04.025.

150 *chronic conditions of food*: Barbara A. Laraia, "Food Insecurity and Chronic Disease," *Advances in Nutrition* 4, no. 2 (2013): 203–12.

151 *a half million people*: Centers for Disease Control and Prevention, "Lyme Disease Data and Surveillance," August 29, 2022, https://www.cdc.gov/lyme/datasurveillance/index.html.

154 *a cornerstone of such efforts*: F. I. Bastos and S. A. Strathdee, "Evaluating Effectiveness of Syringe Exchange Programmes: Current Issues and Future Prospects," *Social Science and Medicine* 51, no. 12 (2000) 1771–82, https://doi.org/10.1016/S0277-9536(00)00109-X.

154 *helped prevent countless deaths*: Theresa Winhusen et al., "Evaluation of a Personally-Tailored Opioid Overdose Prevention Education and Naloxone Distribution Intervention to Promote Harm Reduction and Treatment Readiness in Individuals Actively Using Illicit Opioids," *Drug and Alcohol Dependence* 216 (November 1, 2020): 108265, https://doi.org/10.1016/j.drugalcdep.2020.108265. For further reading, see: Kathryn F. Hawk, Federico E. Vaca, and Gail D'Onofrio, "Reducing Fatal Opioid Overdose: Prevention, Treatment and Harm Reduction Strategies," *Yale Journal of Biology and Medicine* 88, no. 3 (September 3, 2015): 235–45. Also: Su Albert et al., "Project Lazarus: Community-Based Overdose Prevention in Rural North Carolina," *Pain Medicine* 12, no. S2 (2011): S77—S85.

6. THE BODY KEEPS THE STORM

160 *Chimney Tops Trail*: FOX 6 Now Milwaukee, "Charges Dismissed against Boys, Ages 17 and 15 after Gatlinburg Wildfires That Killed More Than a Dozen," July 2, 2017, https://www.fox6now.com/news/charges-dismissed-against-boys-ages-17-and-15-after-gatlinburg-wildfires-that-killed-more-than-a-dozen.

162 *memories of the traumatic event*: Maria L. Pacella, Bryce Hruska, and Douglas L. Delahanty, "The Physical Health Consequences of PTSD and PTSD Symptoms: A Meta-analytic Review," *Journal of Anxiety Disorders* 27, no. 1 (2013): 33–46.

162 *might manifest as difficulty sleeping*: Charles Stewart E. Weston, "Posttraumatic Stress Disorder: A Theoretical Model of the Hyperarousal Subtype," *Frontiers in Psychiatry* 5 (2014): 37.

163 *night in November 2016*: Michael Reed, email correspondence with the author, January 28, 2023. Unless otherwise specified, all Reed quotes in this chapter are from this email correspondence (and that which continued through February 2, 2023).

163 *innocuous stimuli that merely resemble:* Michael Koenigs and Jordan Grafman, "Posttraumatic Stress Disorder: The Role of Medial Prefrontal Cortex and Amygdala," *Neuroscientist* 15, no. 5 (2009): 540–48.

163 *hippocampus can shrink*: Fu Lye Woon, Shabnam Sood, and Dawson W. Hedges, "Hippocampal Volume Deficits Associated with Exposure to Psychological Trauma and Posttraumatic Stress Disorder in Adults: A Meta-analysis," *Progress in Neuro-Psychopharmacology and Biological Psychiatry* 34, no. 7 (2010): 1181–88.

164 *as if they are happening*: Alice Shaam Al Abed et al., "Preventing and Treating PTSD-like Memory by Trauma Contextualization," *Nature Communications* 11, no. 1 (August 24, 2020): 4220, https://doi.org/10.1038/s41467-020-18002-w.

164 *anxiety and hyperarousal*: Lisa M. Shin, Scott L. Rauch, and Roger K. Pitman, "Amygdala, Medial Prefrontal Cortex, and Hippocampal Function in PTSD," *Annals of the New York Academy of Sciences* 1071, no. 1 (2006): 67–79.

164 *cardiovascular system is a prime target:* Updesh Singh Bedi and Rohit Arora, "Cardiovascular Manifestations of Posttraumatic Stress Disorder," *Journal of the National Medical Association* 99, no. 6 (2007): 642.

164 *bears the brunt*: Bianca Augusta Oroian et al., "New Metabolic, Digestive, and Oxidative Stress-Related Manifestations Associated with Posttraumatic Stress Disorder," *Oxidative Medicine and Cellular Longevity* 2021, no. 1 (2021): 5599265, https://doi.org/10.1155/2021/5599265.

165 *can further degrade*: Sian M. J. Hemmings et al., "The Microbiome in Post-traumatic Stress Disorder and Trauma-Exposed Controls: An Exploratory Study," *Psychosomatic Medicine* 79, no. 8 (2017): 936.

165 *immune system isn't immune*: Marpe Bam et al., "Dysregulated Immune System Networks in War Veterans with PTSD Is an Outcome of Altered miRNA Expression and DNA Methylation," *Scientific Reports* 6, no. 1 (2016): 31209.

165 *United States will experience*: National Institute of Mental Health (NIMH), "Post-Traumatic Stress Disorder (PTSD)," accessed August 1, 2023, https://www.nimh.nih.gov/health/statistics/post-traumatic-stress-disorder-ptsd.

166 *in front of the Supreme Court*: Grist Creative, "Why We Must Declare a Global Climate Emergency," Grist, September 25, 2019, https://grist.org/article/why-we-must-declare-a-global-climate-emergency/.

167 *likely suffering from PTSD*: Jean Rhodes et al., "The Impact of Hurricane Katrina on the Mental and Physical Health of Low-Income Parents in New Orleans," *American Journal of Orthopsychiatry* 80, no. 2 (2010): 237.

168 *colleagues called their findings*: Yoko Nomura et al., "Prenatal Exposure to a Natural Disaster and Early Development of Psychiatric Disorders during the Preschool Years: Stress in Pregnancy Study," *Journal of Child Psychology and Psychiatry* 64, no. 7 (2023): 1080–91.

168 *might be passed to offspring*: Ali Jawaid, Martin Roszkowski, and Isabelle M. Mansuy, "Transgenerational Epigenetics of Traumatic Stress," *Progress in Molecular Biology and Translational Science* 158 (2018): 273–98.

168 *reveal that the children*: Yael Danieli, Fran H. Norris, and Brian Engdahl, "A Question of Who, Not If: Psychological Disorders in Holocaust Survivors' Children," *Psychological Trauma: Theory, Research, Practice, and Policy* 9, no. S1 (2017): 98. For further reading, see: Kenneth O'Brien, "The Intergenerational Transference of Post-Traumatic Stress Disorder amongst Children and Grandchildren of Vietnam Veterans in Australia: An Argument for a Genetic Origin," in *Social Change in the 21st Century: 2004 Conference Proceedings*, ed. L. Buys, C. Bailey, and D. Cabrera (Brisbane: Centre for Social Change Research, QUT, 2004), 1–13.

168 *mild shock produced offspring*: Brian G. Dias and Kerry J. Ressler, "Parental Olfactory Experience Influences Behavior and Neural Structure in Subsequent Generations," *Nature Neuroscience* 17, no. 1 (2014): 89–96.

168 *altered stress hormone profiles*: Danieli, Norris, and Engdahl, "A Question of Who, Not If."

170 *to extinguish fear*: Daniel V. Zuj et al., "The Centrality of Fear Extinction in Linking Risk Factors to PTSD: A Narrative Review," *Neuroscience and Biobehavioral Reviews* 69 (2016): 15–35.

172 *determine our feelings and behaviors*: Barbara Olasov Rothbaum et al., "Cognitive-Behavioral Therapy," in *Effective Treatments for PTSD: Practice Guidelines from the International Society for Traumatic Stress Studies*, ed. David Forbes et al. (New York: Guilford Press, 2000), 320–25.

172 *bilateral stimulation of the brain*: Robert Stickgold, "EMDR: A Putative Neurobiological Mechanism of Action," *Journal of Clinical Psychology* 58, no. 1 (2002): 61–75, https://doi.org/10.1002/jclp.1129.

173 *deadliest fire in California*: Cleve R. Wootson Jr., "Camp Fire, California's Deadliest Wildfire in History, Finally Contained," *Washington Post*, November 26, 2018, https://www.washingtonpost.com/nation/2018/11/25/camp-fire -deadliest-wildfire-californias-history-has-been-contained/.

173 *people lost their homes*: Priyanka Boghani, "Camp Fire: By the Numbers," *Frontline*, PBS, October 29, 2019, https://www.pbs.org/wgbh/frontline/arti cle/camp-fire-by-the-numbers/.

173 *indirectly exposed to the fire*: Sarita Silveira et al., "Chronic Mental Health Sequelae of Climate Change Extremes: A Case Study of the Deadliest Californian Wildfire," *International Journal of Environmental Research and Public Health* 18, no. 4 (2021): 1487.

176 *"make you hypervigilant"*: Jyoti Mishra, telephone interview with the author, January 30, 2023. Unless otherwise specified, all Mishra quotes in this chapter are from this interview.

178 *encourage climate-change conversation*: Scott Shigeoka, "The Climate Activist

Who Hasn't Given Up on Mainstream America," Grist, December 23, 2019, https://grist.org/climate/the-climate-activist-who-hasnt-given-up-on-mainstream-america/.

178 *"parts of being a human"*: Anna Jane Joyner, telephone interview with the author, February 1, 2023. Unless otherwise specified, all Joyner quotes in this chapter are from this interview.

178 *phenomenon known as neural coupling*: Greg J. Stephens, Lauren J. Silbert, and Uri Hasson, "Speaker–Listener Neural Coupling Underlies Successful Communication," *Proceedings of the National Academy of Sciences* 107, no. 32 (2010): 14425–30.

181 *"share their vulnerable stories"*: Mary Anne Hitt and Anna Jane Joyner, "The Uses of Sorrow: Anna Jane," November 17, 2021, in *No Place Like Home*, podcast, https://www.spreaker.com/user/15244480/s4-ep-3-the-uses-of-sorrow-anna-jane.

182 *"you turn over and you die"*: Hitt and Joyner, "The Uses of Sorrow: Anna Jane." The remaining Joyner quotes in this chapter come from this interview by Hitt.

7. KARL FRISTON'S THEORY OF EVERYTHING

188 *kinds of future acidic conditions*: Stephen D. Simpson et al., "Ocean Acidification Erodes Crucial Auditory Behaviour in a Marine Fish," *Biology Letters* 7, no. 6 (2011): 917–20.

188 *attracted to smells*: Philip L. Munday et al., "Ocean Acidification Impairs Olfactory Discrimination and Homing Ability of a Marine Fish," *Proceedings of the National Academy of Sciences* 106, no. 6 (February 10, 2009): 1848–52, https://doi.org/10.1073/pnas.0809996106.

189 *"prove to be beneficial or detrimental"*: Sean Bignami et al., "Ocean Acidification Alters the Otoliths of a Pantropical Fish Species with Implications for Sensory Function," *Proceedings of the National Academy of Sciences* 110, no. 18 (April 30, 2013): 7366–70, https://doi.org/10.1073/pnas.1301365110.

189 *appeared to confirm this suspicion*: Göran E. Nilsson et al., "Near-Future Carbon Dioxide Levels Alter Fish Behaviour by Interfering with Neurotransmitter Function," *Nature Climate Change* 2, no. 3 (March 2012): 201–4, https://doi.org/10.1038/nclimate1352.

192 *anecdotes like the following*: David H. Hubel and Torsten N. Wiesel, *Brain and Visual Perception: The Story of a 25-Year Collaboration* (New York: Oxford University Press, 2004).

195 *that's the same thing as saying*: Karl Friston, "The Free-Energy Principle: A Unified Brain Theory?," *Nature Reviews Neuroscience* 11, no. 2 (February 2010): 127–38, https://doi.org/10.1038/nrn2787.

195 *"famously inscrutable"*: John McCrone, "Friston's Theory of Everything," *Lancet Neurology* 21, no. 5 (May 1, 2022): 494, https://doi.org/10.1016/S1474 -4422(22)00137-5. I've liberally lifted McCrone's title for this chapter's title.

196 *a function of water pH*: Angie M. Michaiel and Amy Bernard, "Neurobiology and Changing Ecosystems: Toward Understanding the Impact of Anthropogenic Influences on Neurons and Circuits," *Frontiers in Neural Circuits* 16 (2022): 995354, https://doi.org/10.3389/fncir.2022.995354.

196 *disruptive effects of carbon dioxide*: Christina C. Roggatz et al., "Becoming Nose-Blind—Climate Change Impacts on Chemical Communication," *Global Change Biology* 28, no. 15 (2022): 4495–5055.

196 *impair the homing abilities*: Simone Tosi, Giovanni Burgio, and James C. Nieh, "A Common Neonicotinoid Pesticide, Thiamethoxam, Impairs Honey Bee Flight Ability," *Scientific Reports* 7, no. 1 (April 26, 2017): 1201, https://doi .org/10.1038/s41598-017-01361-8.

197 *"ability to update the model"*: Karl Friston, telephone interview with the author, January 4, 2023. Unless otherwise specified, all Friston quotes in this chapter are from this interview.

197 *how delusions might arise*: Rick A. Adams et al., "Everything Is Connected: Inference and Attractors in Delusions," *Schizophrenia Research* 245 (2022): 5–22.

198 *notion of time's arrow*: Arthur S. Eddington, *The Nature of the Physical World: Gifford Lectures of 1927*, annotated and introduced by H. G. Callaway (Newcastle upon Tyne: Cambridge Scholars Publishing, 2014).

198 *"ultimate equilibrium is death"*: Morten L. Kringelbach and Gustavo Deco, "What Can a Thermodynamics of Mind Say about How to Thrive?," *Aeon*, February 22, 2022, https://aeon.co/essays/what-can-a-thermodynamics-of -mind-say-about-how-to-thrive.

200 *"than when resting"*: Kringelbach and Deco, "What Can a Thermodynamics of Mind Say about How to Thrive?"

202 *"a familiar econiche"*: Adam Linson et al., "The Active Inference Approach to Ecological Perception: General Information Dynamics for Natural and Artificial Embodied Cognition," *Frontiers in Robotics and AI* 5 (2018), https:// www.frontiersin.org/articles/10.3389/frobt.2018.00021.

203 *a curious paper tumbled*: Jérôme Sueur, Bernie Krause, and Almo Farina, "Climate Change Is Breaking Earth's Beat," *Trends in Ecology and Evolution* 34, no. 11 (November 1, 2019): 971–73, https://doi.org/10.1016/j.tree.2019.07.014.

204 *influence reproduction in frogs*: James P. Gibbs and Alvin R. Breisch, "Climate Warming and Calling Phenology of Frogs near Ithaca, New York, 1900– 1999," *Conservation Biology* 15, no. 4 (2001): 1175–78, https://doi.org/10.1046 /j.1523-1739.2001.0150041175.x.

204 *can constrict to minimize*: E.-D. Schulze et al., "Stomatal Responses to Changes in Temperature at Increasing Water Stress," *Planta* 110, no. 1 (March 1, 1973): 29–42, https://doi.org/10.1007/BF00386920.

207 *surprise minimization effectively formalizes*: Karl Friston, "The Variational Principles of Action," in *Geometric and Numerical Foundations of Movements*, ed. Jean-Paul Laumond, Nicolas Mansard, and Jean-Bernard Lasserre, Springer Tracts in Advanced Robotics (Cham: Springer International Publishing, 2017), 117:207–35, https://doi.org/10.1007/978-3-319-51547-2_10.

209 *the so-called Galloping Gertie*: Albert F. Gunns, "The First Tacoma Narrows Bridge: A Brief History of Galloping Gertie," *Pacific Northwest Quarterly* 72, no. 4 (1981): 162–69.

210 *"make them see things differently"*: Alejandro de la Garza, "Climate Protesters Are Throwing Soup at Art. A Brooklyn Psychologist Is Behind It," *Time*, November 18, 2022, https://time.com/6234840/art-climate-protests-margaret-klein-salamon/.

211 *"didn't understand it at first"*: Timothy Morton, "You Are Ecological," Brainwash Festival 2022, November 10, 2022, https://www.youtube.com/watch?v=dGTQS6_SBdc.

8. BURN SCAR

213 *An artist's diary*: Zuzanna Stańska, "The Mysterious Street from Edvard Munch's The Scream," *DailyArt Magazine* (blog), May 18, 2023, https://www.dailyartmagazine.com/the-mysterious-road-of-the-scream-by-edvard-munch/.

214 *friend of Munch's recalled*: Richard Panek, "ART; 'The Scream,' East of Krakatoa," *New York Times*, February 8, 2004, sec. Arts, https://www.nytimes.com/2004/02/08/arts/art-the-scream-east-of-krakatoa.html.

215 *great Krakatoa eruption*: Donald W. Olson, Russell L. Doescher, and Marilynn S. Olson, "When the Sky Ran Red: The Story behind *The Scream*," *Sky and Telescope*, February 2004, 28.

217 *example of climate concern*: Andreea Bratu et al., "The 2021 Western North American Heat Dome Increased Climate Change Anxiety among British Columbians: Results from a Natural Experiment," *Journal of Climate Change and Health* 6 (2022): 100116.

218 *"the mountain is gone"*: Kathy Selvage, telephone interview with the author, January 21, 2016. Unless otherwise specified, all Selvage quotes in this chapter are from this interview.

219 *covers the niche field*: Michael Hendryx and Kestrel A. Innes-Wimsatt, "Increased Risk of Depression for People Living in Coal Mining Areas of Central Appalachia," *Ecopsychology* 5, no. 3 (September 2013): 179–87, https://doi.org

/10.1089/eco.2013.0029; and Paige Cordial, Ruth Riding-Malon, and Hilary Lips, "The Effects of Mountaintop Removal Coal Mining on Mental Health, Well-Being, and Community Health in Central Appalachia," *Ecopsychology* 4, no. 3 (2012): 201–8.

220 *"biological issues make more sense"*: Michael Hendryx, telephone interview with the author, November 5, 2015. Unless otherwise specified, all Hendryx quotes in this chapter are from this interview.

220 *pain and yearning*: Glenn Albrecht, "'Solastalgia.' A New Concept in Health and Identity," *PAN: Philosophy Activism Nature* 3 (2005): 41–55.

221 *"changed around you"*: Paige Cordial, telephone interview with the author, November 5, 2015. Unless otherwise specified, all Cordial quotes in this chapter are from this interview.

221 *"acute and severe enough"*: Paige Cordial, "A Qualitative Exploration of the Effects of Mountaintop Removal on the Wellness of Central Appalachians Living near Surface Mines" (PhD diss., Radford University, May 2013), http://wagner.radford.edu/115/3/Paige%20Cordial%20Final%20Dissertation%205-3-13.pdf.

221 *in a 2011 interview*: Jeff Biggers, "Next Steps for Anti-Mountaintop Removal Movement: Interview with Appalachian Leader Bo Webb," *Indypendent*, March 29, 2011, https://indypendent.org/2011/03/next-steps-for-anti-mountaintop-removal-movement-interview-with-appalachian-leader-bo-webb/.

221 *"walking on the moon"*: Cordial, "A Qualitative Exploration of the Effects of Mountaintop Removal."

222 *"intervene directly in human thought"*: Amitav Ghosh, *The Great Derangement: Climate Change and the Unthinkable* (New York: Penguin Books, 2016).

222 *encompass three distinct processes*: Alissa J. Mrazek, Tokiko Harada, and Joan Y. Chiao, "Cultural Neuroscience of Identity Development," in *The Oxford Handbook of Identity Development*, ed. Kate C. McLean and Moin Syed (New York: Oxford University Press, 2014), 423.

223 *performing the same action*: Giuseppe Di Pellegrino et al., "Understanding Motor Events: A Neurophysiological Study," *Experimental Brain Research* 91 (1992): 176–80.

223 *reflect and help shape our social interactions*: Luca Bonini et al., "Mirror Neurons 30 Years Later: Implications and Applications," *Trends in Cognitive Sciences* 26, no. 9 (September 1, 2022): 767–81, https://doi.org/10.1016/j.tics.2022.06.003.

223 *suggest a more direct route*: Vittorio Gallese, "Mirror Neurons, Embodied Simulation, and the Neural Basis of Social Identification," *Psychoanalytic Dialogues* 19, no. 5 (2009): 519–36.

225 *photographer Roger May*: Becky Harlan, "A Fresh Look at Appalachia—50 Years after the War on Poverty," *National Geographic*, February 6, 2015, https://www.nationalgeographic.com/photography/article/a-fresh-look-at-appalachia-50-years-after-the-war-on-poverty.

225 *the logic of environmental degradation*: Holly Vins et al., "The Mental Health Outcomes of Drought: A Systematic Review and Causal Process Diagram," *International Journal of Environmental Research and Public Health* 12, no. 10 (October 2015): 13251–75, https://doi.org/10.3390/ijerph121013251.

226 *shortage of mental health care workers*: Michael Hendryx, "Mental Health Professional Shortage Areas in Rural Appalachia," *Journal of Rural Health* 24, no. 2 (2008): 179–82, https://doi.org/10.1111/j.1748-0361.2008.00155.x.

226 *A 2009 nationwide study*: Kathleen C. Thomas et al., "County-Level Estimates of Mental Health Professional Shortage in the United States," *Psychiatric Services* 60, no. 10 (2009): 1323–28.

227 *people in poor mental health*: Hendryx, "Mental Health Professional Shortage Areas in Rural Appalachia."

229 *"breathe in these toxins"*: Ciara O'Rourke, "Climate Change's Hidden Victim: Your Mental Health," *OneZero*, January 24, 2019, https://onezero.medium.com/the-emotional-damage-done-by-climate-change-2f8f9ad59155.

229 *state's worst fire season*: Blacki Migliozzi et al., "Record Wildfires on the West Coast Are Capping a Disastrous Decade," *New York Times*, September 24, 2020, sec. Climate, https://www.nytimes.com/interactive/2020/09/24/climate/fires-worst-year-california-oregon-washington.html.

230 *appears to have slowed down*: Chess Stetson, Matthew P. Fiesta, and David M. Eagleman, "Does Time Really Slow Down during a Frightening Event?," *PLOS ONE* 2, no. 12 (2007): e1295.

230 *as Zadie Smith wrote*: Zadie Smith, "Elegy for a Country's Seasons," *New York Review of Books* 61, no. 6 (2014): 6.

231 *"very nature demands attention"*: Abdul-Ghaaliq Lalkhen, *An Anatomy of Pain: How the Body and the Mind Experience and Endure Physical Suffering* (New York: Simon and Schuster, 2022).

231 *takes ages, electrically speaking*: Charlotte E. Steeds, "The Anatomy and Physiology of Pain," *Surgery* (Oxford) 27, no. 12 (2009): 507–11.

231 *a powerful communicator*: Harald Breivik et al., "Assessment of Pain," *British Journal of Anaesthesia* 101, no. 1 (2008): 17–24. For further reading, see: Kenneth D. Craig, "The Social Communication Model of Pain," *Canadian Psychology/Psychologie canadienne* 50, no. 1 (2009): 22. Also: Simon W. Townsend et al., "Flexible Alarm Calling in Meerkats: The Role of the Social Environment and

Predation Urgency," *Behavioral Ecology* 23, no. 6 (November 1, 2012): 1360–64, https://doi.org/10.1093/beheco/ars129.

234 *the practice of shinrin-yoku*: Yasuhiro Kotera, Miles Richardson, and David Sheffield, "Effects of Shinrin-Yoku (Forest Bathing) and Nature Therapy on Mental Health: A Systematic Review and Meta-analysis," *International Journal of Mental Health and Addiction* (2020): 1–25.

236 *its neural branches withering*: F. Ohl et al., "Effect of Chronic Psychosocial Stress and Long-Term Cortisol Treatment on Hippocampus-Mediated Memory and Hippocampal Volume: A Pilot-Study in Tree Shrews," *Psychoneuroendocrinology* 25, no. 4 (May 2000): 357–63, https://doi.org/10.1016/S0306-4530(99)00062-1.

236 *susceptible to the eroding*: Jessica M. McKlveen et al., "Chronic Stress Increases Prefrontal Inhibition: A Mechanism for Stress-Induced Prefrontal Dysfunction," *Biological Psychiatry* 80, no. 10 (2016): 754–64.

9. THE GRAMMAR OF EARTH

240 *distinguish between various hues*: Jonathan Winawer et al., "Russian Blues Reveal Effects of Language on Color Discrimination," *Proceedings of the National Academy of Sciences* 104, no. 19 (May 8, 2007): 7780–85, https://doi.org/10.1073/pnas.0701644104.

240 *discern the subtleties*: Julie Goldstein, Jules Davidoff, and Debi Roberson, "Knowing Color Terms Enhances Recognition: Further Evidence from English and Himba," *Journal of Experimental Child Psychology* 102, no. 2 (2009): 219–38.

240 *memory and other cognitive processes*: Martin Maier and Rasha Abdel Rahman, "Native Language Promotes Access to Visual Consciousness," *Psychological Science* 29, no. 11 (2018): 1757–72. For further reading, see: Jasna Martinovic, Galina V. Paramei, and W. Joseph MacInnes, "Russian Blues Reveal the Limits of Language Influencing Colour Discrimination," *Cognition* 201 (August 1, 2020): 104281, https://doi.org/10.1016/j.cognition.2020.104281.

241 *a distinct temporal framework*: Benjamin Lee Whorf, *Language, Thought, and Reality: Selected Writings of Benjamin Lee Whorf* (Cambridge, MA: MIT Press, 2012).

242 *progression of time is in accordance*: Lera Boroditsky and Alice Gaby, "Remembrances of Times East: Absolute Spatial Representations of Time in an Australian Aboriginal Community," *Psychological Science* 21, no. 11 (2010): 1635–39.

243 *expected to survive:* John Noble Wilford, "Languages Die, but Not Their Last Words," *New York Times*, September 19, 2007, https://www.nytimes.com/2007/09/19/science/19language.html.

243 *climatic and biodiversity factors*: L. J. Gorenflo et al., "Co-occurrence of Linguistic and Biological Diversity in Biodiversity Hotspots and High Biodiversity Wilderness Areas," *Proceedings of the National Academy of Sciences* 109, no. 21 (May 22, 2012): 8032–37, https://doi.org/10.1073/pnas.1117511109. For further reading, see: Lenore A. Grenoble, "Arctic Indigenous Languages: Vitality and Revitalization," in *The Routledge Handbook of Language Revitalization*, ed. Leanne Hinton, Leena Huss, and Gerald Roche (New York: Routledge, 2018). Also: Gary Paul Nabhan, Patrick Pynes, and Tony Joe, "Safeguarding Species, Languages, and Cultures in the Time of Diversity Loss: From the Colorado Plateau to Global Hotspots," *Annals of the Missouri Botanical Garden* 89, no. 2 (2002): 164–75, https://doi.org/10.2307/3298561.

244 *"agency in people shifting"*: Mandana Seyfeddinipur, telephone interview with the author, February 3, 2023. Unless otherwise specified, all Seyfeddinipur quotes in this chapter are from this interview.

247 *the classic language network*: Gregory Hickok, "The Functional Neuroanatomy of Language," *Physics of Life Reviews* 6, no. 3 (2009): 121–43.

248 *serves as a pathway connecting*: Hickok, "Functional Neuroanatomy of Language."

248 *dynamic nature of language representation*: Ellen Bialystok and Xiaojia Feng, "Language Proficiency and Executive Control in Proactive Interference: Evidence from Monolingual and Bilingual Children and Adults," *Brain and Language* 109, nos. 2–3 (2009): 93–100. For further reading, see: Ellen Bialystok and Mythili Viswanathan, "Components of Executive Control with Advantages for Bilingual Children in Two Cultures," *Cognition* 112, no. 3 (2009): 494–500.

250 *intensive burst of air*: Caleb Everett, "Evidence for Direct Geographic Influences on Linguistic Sounds: The Case of Ejectives," *PLOS ONE* 8, no. 6 (June 12, 2013): e65275, https://doi.org/10.1371/journal.pone.0065275.

250 *proximity to rich biodiversity*: Benjamin T. Wilder et al., "The Importance of Indigenous Knowledge in Curbing the Loss of Language and Biodiversity," *BioScience* 66, no. 6 (2016): 499–509. For further reading, see: Gorenflo et al., "Co-occurrence of Linguistic and Biological Diversity."

250 *compass directions for spatial orientations*: Stephen C. Levinson, "Language and Cognition: The Cognitive Consequences of Spatial Description in Guugu Yimithirr," *Journal of Linguistic Anthropology* 7, no. 1 (1997): 98–131.

250 *London cab drivers*: Eleanor A. Maguire, Katherine Woollett, and Hugo J. Spiers, "London Taxi Drivers and Bus Drivers: A Structural MRI and Neuropsychological Analysis," *Hippocampus* 16, no. 12 (2006): 1091–1101.

251 *languages enhances cognitive flexibility*: Judith F. Kroll and Paola E. Dussias,

"The Benefits of Multilingualism to the Personal and Professional Development of Residents of the US," *Foreign Language Annals* 50, no. 2 (2017): 248–59.

251 *delay the onset of diseases*: Hilary D. Duncan et al., "Structural Brain Differences between Monolingual and Multilingual Patients with Mild Cognitive Impairment and Alzheimer Disease: Evidence for Cognitive Reserve," *Neuropsychologia* 109 (2018): 270–82. For further reading, see: Howard Chertkow et al., "Multilingualism (but Not Always Bilingualism) Delays the Onset of Alzheimer Disease: Evidence from a Bilingual Community," *Alzheimer Disease and Associated Disorders* 24, no. 2 (2010): 118–25.

251 *substantial gray-matter plasticity*: Stefan Elmer, Jürgen Hänggi, and Lutz Jäncke, "Processing Demands upon Cognitive, Linguistic, and Articulatory Functions Promote Grey Matter Plasticity in the Adult Multilingual Brain: Insights from Simultaneous Interpreters," *Cortex* 54 (2014): 179–89.

251 *"brain manages this collision"*: Anouschka Foltz, telephone interview with the author, February 3, 2023. Unless otherwise specified, all Foltz quotes in this chapter are from this interview.

254 *coldest town in mainland Norway*: Reece Toth, "The Coldest Places in Europe: 7 Ultimate Frigid Towns," Travel Snippet, March 21, 2023, https://travelsnippet .com/europe/coldest-places-in-europe/.

255 *"the rain freezes as plates"*: Jonna Utsi, interview with the author, Guovdageaidnu, Norway, February 7, 2023. Unless otherwise specified, all Utsi quotes in this chapter are from this interview.

257 *"what I'd planned to do"*: Annika Pasanen, interview with the author, Guovdageaidnu, Norway, January 31, 2023. Unless otherwise specified, all Pasanen quotes in this chapter are from this interview.

260 *soon began to stir*: William H. Wilson and Kauanoe Kamanā, "'*Mai Loko Mai O Ka 'I'ini*: Proceeding from a Dream': The 'Aha Pūnana Leo Connection in Hawaiian Language Revitalization," in *The Green Book of Language Revitalization in Practice*, ed. Leanne Hinton and Kenneth Hale (Leiden: Brill, 2001), 147–76. For further reading, see: Keiki K. C. Kawai'ae'a, Alohalani Kaluhiokalani Housman, and Makalapua Alencastre, "Pū'ā i ka 'Ōlelo, Ola ka 'Ohana: Three Generations of Hawaiian Language Revitalization," *Hulili: Multidisciplinary Research on Hawaiian Well-Being* 4, no. 1 (2007): 183–237.

261 *Welsh-language education*: Colin H. Williams, "The Lightening Veil: Language Revitalization in Wales," *Review of Research in Education* 38, no. 1 (2014): 242–72. For further reading, see: Dylan V. Jones and Marilyn Martin-Jones, "Bilingual Education and Language Revitalization in Wales: Past Achievements and Current Issues," in *Medium of Instruction Policies: Which Agenda? Whose*

Agenda, ed. James W. Tollefson and Amy B. M. Tsui (Mahwah, NJ: Lawrence Erlbaum Associates, 2004), 43–70.

261 *danger of falling silent*: Jeanette King, "Te Kōhanga Reo: Māori Language Revitalization," in Hinton and Hale, *The Green Book of Language Revitalization in Practice*, 119–31. For further reading, see: Kimai Tocker, "The Origins of Kura Kaupapa Māori," *New Zealand Journal of Educational Studies* 50 (2015): 23–38.

INDEX

INDEX

ABOUT THE AUTHOR

CLAYTON PAGE ALDERN is a neuroscientist turned environmental journalist whose work has appeared in *The Atlantic, The Guardian, The New Republic, Mother Jones, Vox, Newsweek, The Economist, Scientific American*, and Grist, where he is a senior data reporter. His climate change data visualizations have appeared in a variety of forums, including on the US Senate floor in a speech by Senator Sheldon Whitehouse.

A Rhodes Scholar and a Reynolds Journalism Institute Fellow, he holds a master's in neuroscience and a master's in public policy from the University of Oxford. He is also a research affiliate at the Center for Studies in Demography and Ecology at the University of Washington and a grantee of the Pulitzer Center. He has contributed to reporting teams that have won a national Edward R. Murrow Award, multiple Online Journalism Awards, and the Breaking Barriers Award from the Institute for Nonprofit News.